OIL & GAS

THE PRODUCTION STORY

BY
Ron Baker

A PETEX PRIMER

Published by

PETROLEUM EXTENSION SERVICE
Division of Continuing Education
The University of Texas at Austin
Austin, Texas

1983

Published 1983. Sixth Impression 2004
Printed in the United States of America

Brand names, company names, trademarks, or other identifying symbols appearing in illustrations or text are used for educational purposes only and do not constitute an endorsement by the author or publisher.

Catalog No. 3.9000
ISBN 0-88698-002-X

The University of Texas at Austin is an equal opportunity employer. No state funds were used to produce this manual.

Contents

Preface

This is a story about oil and gas production. It is for people who have little or no knowledge of what it takes to get oil and gas from their point of origin to the surface where they can be utilized. The production of oil and gas involves a number of processes, techniques, and equipment—on the surface and in the subsurface. Although this story is relatively complex, much effort has been put into keeping it as clear as possible.

The purpose of this story is twofold: to enlighten those whose jobs require a general knowledge of production operations or those who simply wish to be better informed; and to serve as an introduction to the PETEX Oil and Gas Production Series. Publications in this series cover in detail the most common and important aspects of this vital phase of the petroleum industry. The series, published in cooperation with the American Petroleum Institute, is recommended for those who want to know more about production.

In any basic story, accuracy, as well as clarity, is essential. Therefore, a great deal of effort was put forth to ensure correct and up-to-date information. However, readers who discover inaccuracies or inconsistencies are encouraged to bring them to our attention; indeed, suggestions of any kind that may make this book more useful are welcomed.

Ron Baker
Petroleum Extension Service

Acknowledgments

We gratefully acknowledge the technical review provided by the following select members of the Advisory Committee for the API-PETEX School of Production Technology:

M. C. Boyd, Sun Exploration and Production Company, Abilene, Texas;

R. G. Sanders, Gulf Oil Exploration and Production Company, Kilgore, Texas;

S. H. Stocker, Amoco Production Company (USA), Houston, Texas; and

J. C. West, Cypress Energy Development Company, Houston, Texas.

Their comments and guidance contributed much to the final product.

Introduction

Practically all of us use oil and gas—in our automobiles, at home, and at work. Indeed, almost every activity we engage in involves either direct or indirect use of oil or gas. However, in spite of our almost universal use of petroleum, many of us do not know too much about it. Since oil touches all our lives, we really ought to be better informed about it.

Becoming better informed seems difficult. The oil industry has its own language, its own special breed of people who work in it, and its own mystery—oil and gas occur underneath the ground, usually thousands of feet below the surface and well out of sight and mind. Further, what happens on the surface, and a lot does, looks very complicated and difficult to understand. Yet, in reality, the oil industry is not all that complex. Certainly, the basic principles are easy to understand, once they're explained.

In this book, we'll describe many of the basic things necessary to produce oil and gas. The story begins with an explanation of how oil and gas might have originated and how they occur today. Since it takes a well to get oil and gas out of the ground, a little about how wells are drilled and how they are set up to produce is included. What causes oil and gas to flow from the well and what can be done to get oil that won't flow on its own into and up the well are major parts of this story. How oil and gas are handled once on the surface—how they are separated from each other and how further treatment is usually needed before they can be sold to a pipeline company or other shipper—must also be told. Finally, we'll tell about some of the special problems that occur and how they are handled.

Origin and Accumulation of Oil and Gas

An all too popular notion is that oil can be found in underground rivers or lakes. The idea seems to be that large, subterranean caves are filled with oil or that buried rivers flow with pure crude from bank to bank. Nothing could be further from the truth.

Yet, it is easy to understand how such concepts come about. Even experienced oilmen often refer to an *oil pool,* which is a loose term for an underground place where oil is stored. And to further compound the problem, one of the correct names for such a place is *reservoir.* Since many cities store their drinking water in ponds or lakes called reservoirs, this term also adds to the confusion. So, before going any further, let's find out where oil and gas really occur and how they might have gotten there.

Characteristics of Rocks

To understand how oil and gas originated, it is necessary to learn something about the basic characteristics of rocks. Most oil is found in rocks, and that is why oil is called *petroleum.* The word *petroleum* comes from the old Greek word for rock, *petra,* and the Latin word for oil, *oleum.* Thus, petroleum literally means rock oil.

Natural gas is also found in rocks and is often associated with oil. Both oil and natural gas are called *hydrocarbons.* They are called hydrocarbons because they are made up of the two elements, hydrogen and carbon.

It may be a little hard to see how hydrocarbons such as oil and gas can come out of a rock. Most of us think that there are very few things more solid than a rock. But some rocks are not as solid as they appear to the naked eye. Certain rocks contain many tiny openings that can be seen under a strong magnifying glass. These openings are called *pores* or *pore spaces* (fig. 1).

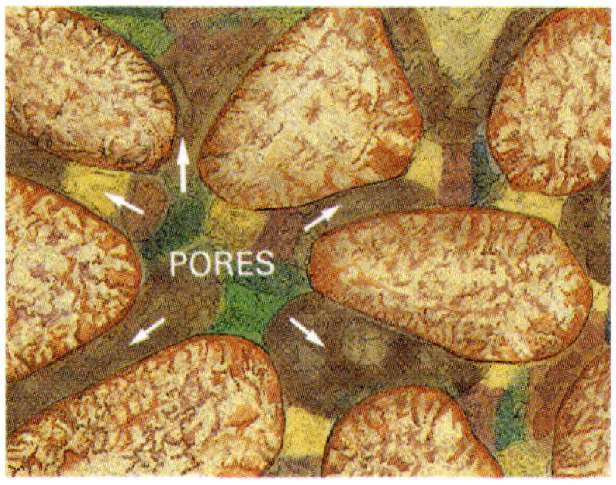

Figure 1. When magnified, the pores in a rock are visible.

If a rock has pores, then it is *porous* and has *porosity.* Any oil and gas in the rock has to be in the pores. Some rocks are quite porous and are said to have *high porosity.* Other rocks do not have many pores or have very small pores and are said to have *low porosity.* But whether the rock has high or low porosity, any oil or natural gas that exists will occur in the pores.

Another desirable feature of oil- and gas-bearing rocks is *permeability.* A rock is permeable, or has permeability, when the pores of the rock are connected (fig. 2).

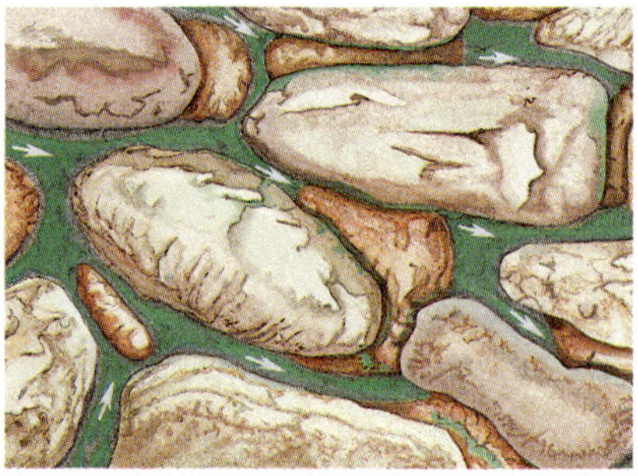
Figure 2. A rock is permeable when the pores are connected.

Permeability provides a way for any oil and gas in the rock to move or flow from pore to pore. As with porosity, some rocks have high permeability—making it easy for fluids like oil and gas to move from one pore to another. Other rocks have low permeability, and fluids have difficulty in moving about in the rock.

When a rock has pores with oil and gas in them, and when the rock is permeable, the situation is promising. It may be possible to drill a well down to the rock and produce oil and gas from it. The oil or gas can flow out of the pores through the permeable channels in the rock and into the well.

Incidentally, some petroleum-bearing rocks are porous but not permeable. Found in the western United States, for example, are vast deposits of a rock that is porous and has hydrocarbons but practically no permeability. This rock is referred to as *shale;* therefore, the hydrocarbons contained in the rock are often called *shale oil.* Because the rock is not permeable, its hydrocarbons cannot be produced through wells. The hydrocarbons have no way to flow out of the rock pores and into a wellbore. Thus, advanced mining techniques and other innovative technology have to be used to get hydrocarbons out of this type of rock.

Origin of Oil and Gas

Many theories exist about how oil and gas were formed. Perhaps the best theory is one called the *organic theory.* The organic theory holds that oil and gas came from the remains of plants and animals. (The organic theory is called *organic* because plants and animals are organisms.) Most geologists think that the plants and animals that gave rise to oil and gas were very small, even microscopic, in size, although very large plants and animals like trees and dinosaurs may have been living at the time petroleum was formed. These tiny plants and animals lived in the rivers and seas that must have existed in Earth's distant past, millions and millions of years ago, even as they do today (fig. 3).

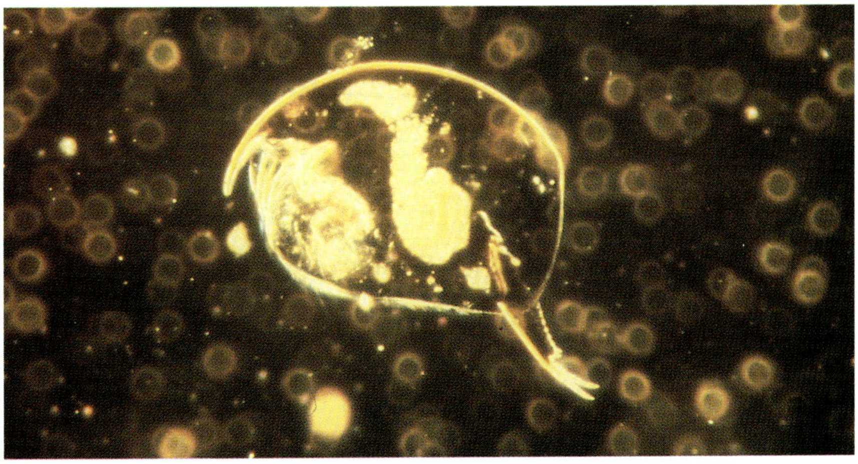

Figure 3. Microscopic animals such as these probably lived in ancient seas.

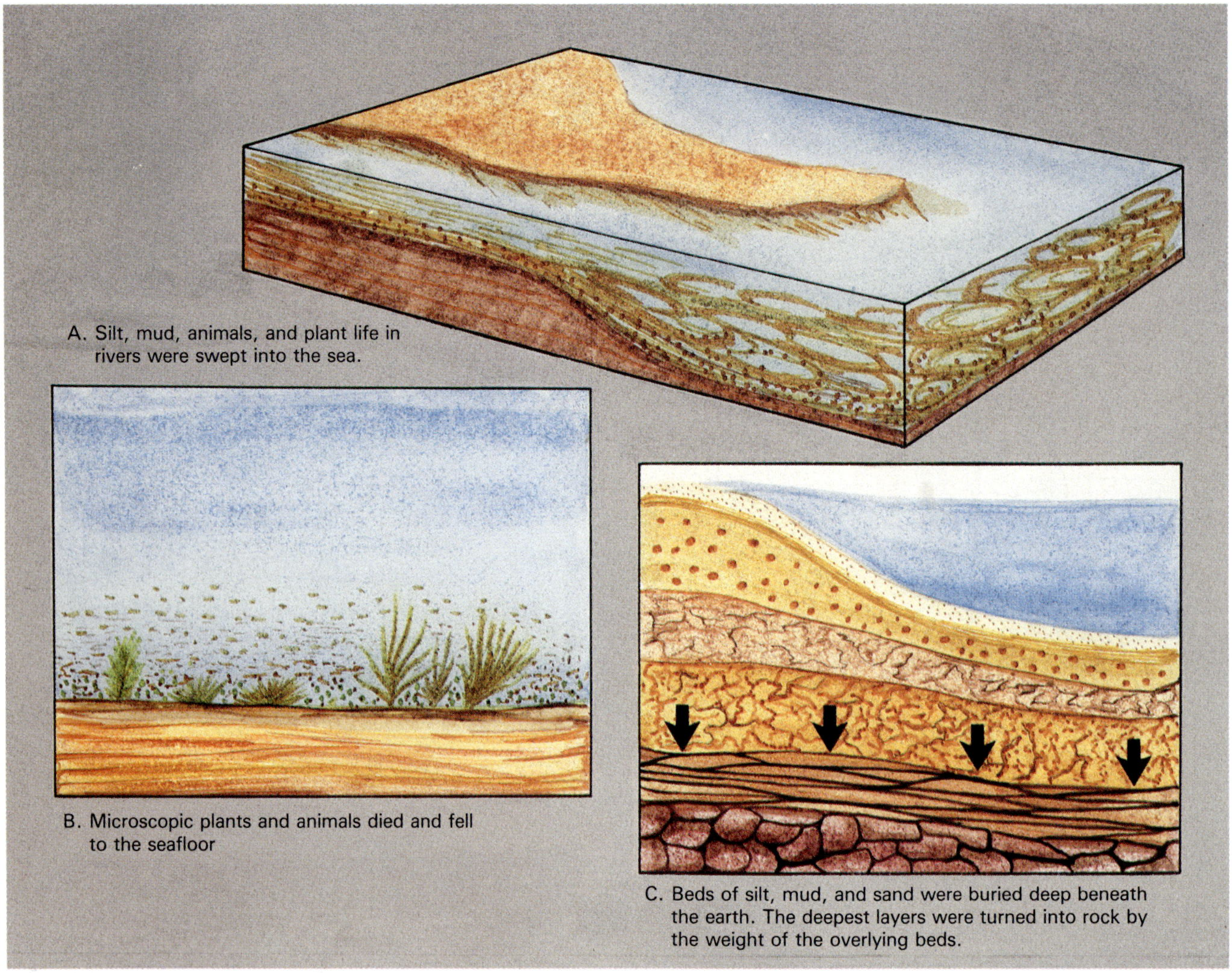

Figure 4. The organic theory holds that oil and gas were formed from the remains of plants and animals.

The plants and animals in the rivers were carried down to the sea, along with river silts and muds (fig. 4*A*). Also, the sea itself was full of microscopic plant and animal life (fig. 4*B*). As these small organisms died, they fell to the ocean floor. On the bottom, they mixed with the silt, sand, and mud. As a result, a rich, organic mixture was formed that was cut off from any oxygen dissolved in the water. With no oxygen, the organic material could not decay as it normally would.

As time went by, day after day for thousands and thousands of years, the silt, sand, mud, and organic material continued to build up. Eventually, a thick body of sediments, enriched by organic remains, accumulated on the bottom of the sea.

More time passed. Sediments continued to be deposited until thick beds of sand and mud lay above the body of sediments containing the organic remains. The great weight of the overlying sediments pushed the lower sediments deep into the earth and caused changes to occur in them. The weight caused the bottom beds to turn into rock (fig. 4*C*). Also, high heat and pressure began to work on the organic substances locked in the lowest layer. Bacteria, chemical reactions, and other forces were at work as well. The end result was that the organic remains were transformed into oil and gas.

Meanwhile, other great forces were at work, forces that would change the face of the earth.

Earthquakes moved blocks of earth upward, downward, and sideways (fig. 5). Layers of rock were folded into arches and troughs (fig. 6). Erosion from wind, rain, and ice wore away the land. The newly formed oil and gas were also affected by these forces—they were moved out of their birthplace.

Great pressures created by the overlying rock beds forced the newly formed petroleum to flow. It flowed through cracks and fissures in the rock and it oozed from pore to pore, constantly moving in a tortuous path, seeking the surface of the earth (fig. 7). Sometimes it actually reached the surface.

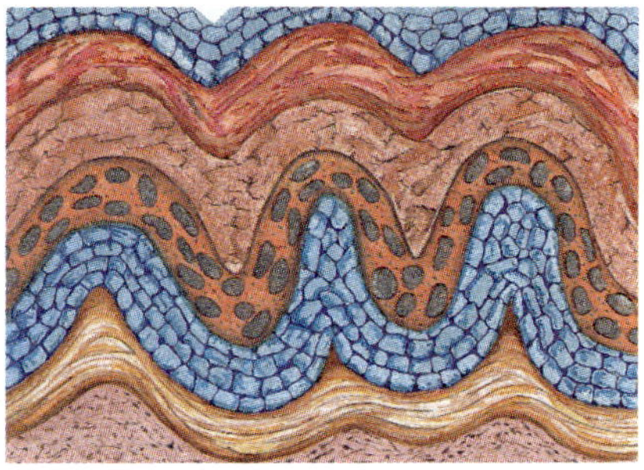

Figure 6. Layers of rock were folded into arches and troughs.

Sometimes its upward movement was blocked by an impermeable rock layer deep below the surface. The oil and gas trapped below the surface are what we drill for today.

Figure 7. Once formed, petroleum was forced to move upward through tiny fissures or pore spaces in the rock.

Figure 5. Earthquakes moved huge blocks of earth up, down, and sideways.

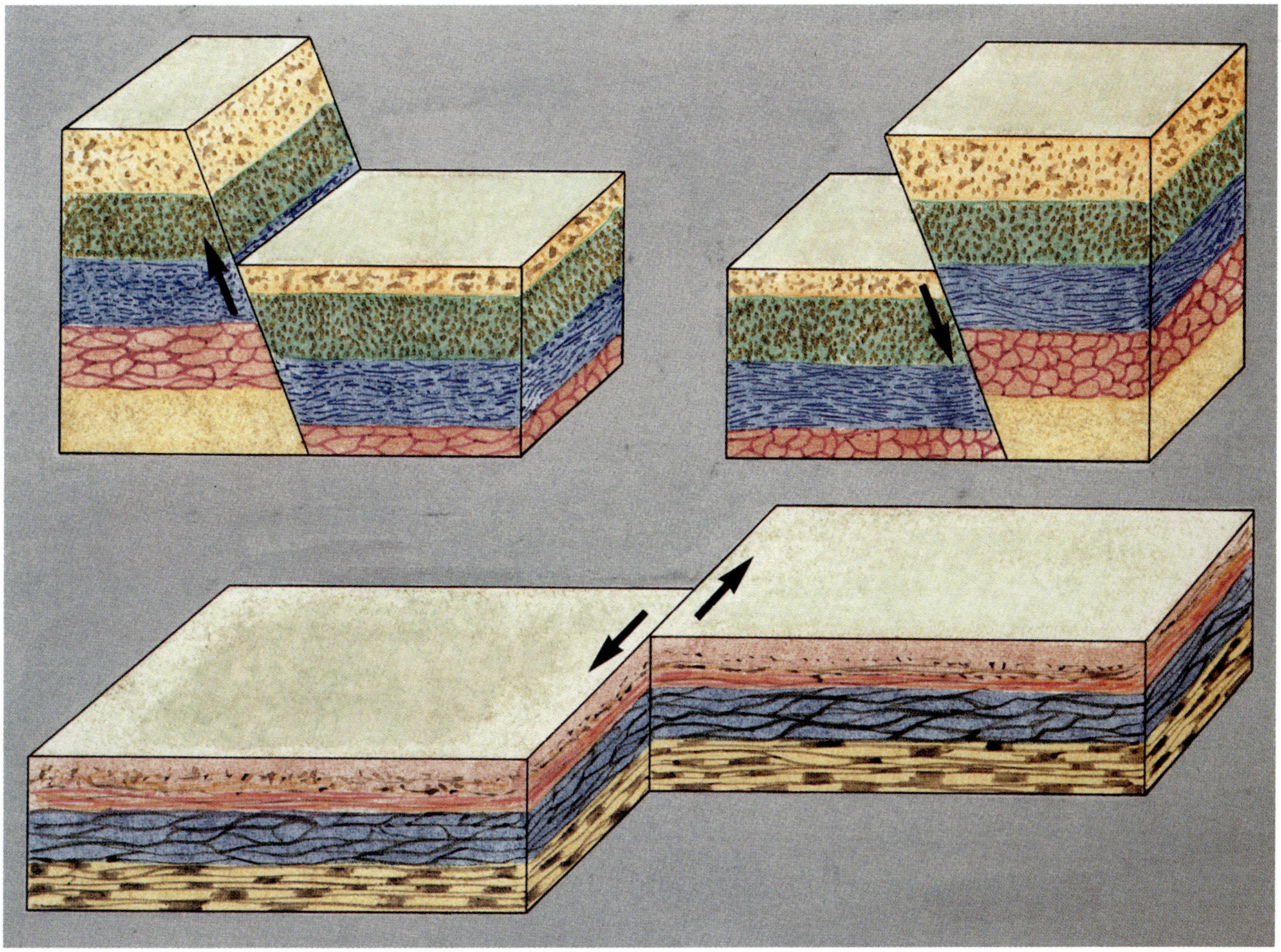

Traps

A petroleum trap is an arrangement of rock layers that contains an accumulation of hydrocarbons. The arrangement consists of an impermeable caprock overlaying porous and permeable rock. Hydrocarbons are stored in the porous layer, and the caprock prevents their escape. The rock that contains the hydrocarbons is called a *reservoir.*

Hydrocarbon traps come in all sizes, shapes, and types. Maybe the easiest way to study them is to look at the geologic features that caused them to occur. The basic kinds of traps are those that occurred because of folding, faulting, unconformities, domes or plugs, changes in permeability within a formation, and combinations of two or more of these.

Traps that were formed because of folding include *anticlines* (fig. 8). In an anticlinal trap, the rock layers that were originally laid down horizontally were folded upward into an arch similar to one over a doorway or window. After the hydrocarbons were formed in their source bed, they migrated into one of the porous and permeable layers in the anticline. The oil and gas then accumulated in the top of the folded, porous layer and were sealed there by an impermeable bed above.

A *fault* is a fracture, or break, in rock layers. It is caused by earth movements. Sometimes, the tilted layers on one side of a fault have moved in relation to the layers on the other side. In fault traps, the reservoir rock is on one side of the fault. On the other side is an impermeable layer that was moved opposite the reservoir and serves to prevent the further movement of hydrocarbons. In other words, the hydrocarbons migrated into the reservoir rock and were trapped there by the impermeable layer (fig. 9).

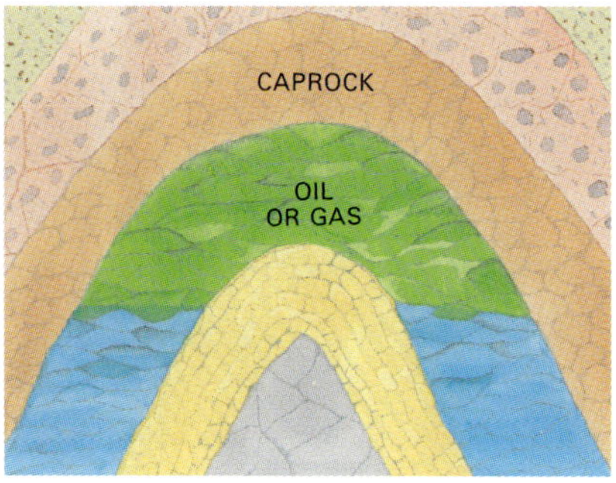

Figure 8. An anticline can trap oil or gas that has migrated into a porous and permeable rock layer. The impermeable caprock prevents the escape of the petroleum.

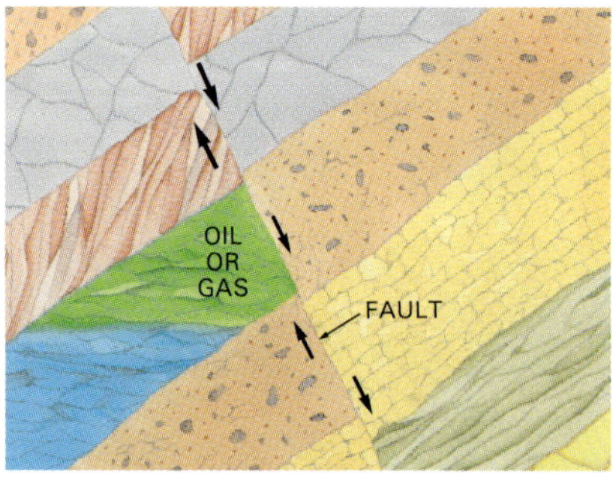

Figure 9. In a fault trap, the layers on one side of the fault have moved opposite the petroleum-bearing layer on the other side.

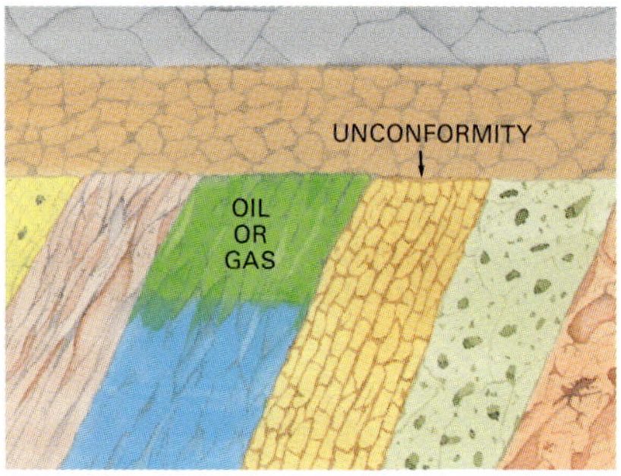

Figure 10. An unconformity can form a trap for petroleum.

Another type of trap occurs because of an *unconformity.* The upward escape of hydrocarbons is stopped by impermeable rocks laid down on top of the weathered surface of the lower beds (fig. 10). An unconformity is a lack of continuity between rock beds, or strata.

To understand, think of several layers of rock being deposited over a very long period of time at the bottom of a sea. Eventually, for some reason—perhaps the sea retreated or evaporated—the rocks stopped being deposited. As more time passed, the layers were gradually tilted by earth movements and, since they were now exposed to the atmosphere, the erosive forces of wind, water, or ice weathered them.

As still more time passed (we're talking about millions of years here), another change occurred in the environment so that new beds were deposited on top of the old, weathered ones. The final result, what we see today, is a series of beds, all in contact with one another, but with a gap in the sequence of layers. The gap occurs because, during a period of time in the history of the structure, no rocks were deposited.

Oil and gas are also found associated with *domes* or *plugs* of material, usually salt, that have pierced and deformed overlying strata (fig. 11). As the salt pushed

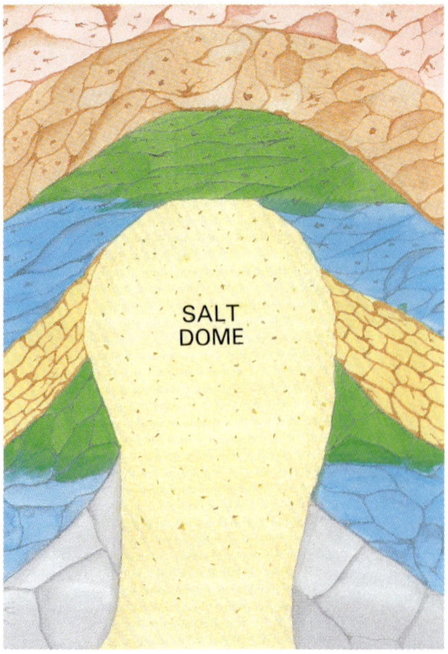

Figure 11. Traps associated with salt domes usually occur on both sides of the dome and sometimes above it.

through the rock layers, it tilted those layers that it pierced, and it folded those above that it did not pierce. Hydrocarbons could migrate into any porous and permeable beds on both sides of the column of salt and be trapped there since salt is impervious. Any folded, porous, and permeable beds above the salt column could also serve as traps for hydrocarbons if overlain by an impermeable rock layer.

Another type of trap was caused by a *change of permeability* within a rock layer (fig. 12). Hydrocarbons migrated into permeable sections of the bed and were trapped there because they were surrounded by impermeable sections of the same bed.

Probably the most common type of reservoir is one that was formed by a *combination* of folding, faulting, changes in permeability, and other things. For example, a reservoir could occur in an anticline that is faulted and associated with an unconformity (fig. 13).

Figure 12. A change of permeability within a rock layer can form a trap. The lower part of the layer is porous and permeable, but the upper part is not.

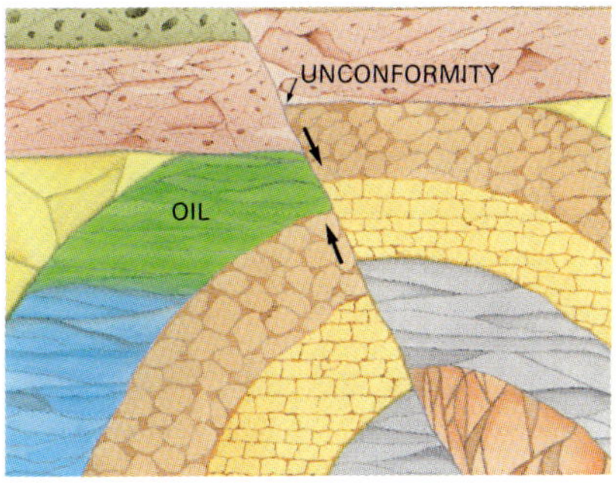

Figure 13. A combination of a folded and faulted trap in association with an unconformity can trap hydrocarbons.

Reservoir Fluids

In all types of traps, oil and gas seldom fill up all the rock in the reservoir. Most of the time, salt water takes up the rest of the space. Much of the salt water lies underneath the oil and gas (fig. 14), but salt water usually coats the grains of the rock where the oil and gas occur as well (fig. 15). This salt water is the ancient seawater that existed when the formation was originally laid down. When hydrocarbons move into a reservoir, they have to move or displace the salt water that is already in the rock pores. But they cannot displace all of the water. In fact, pore spaces can have anywhere from 10 to 50 percent salt water even in the midst of oil and gas.

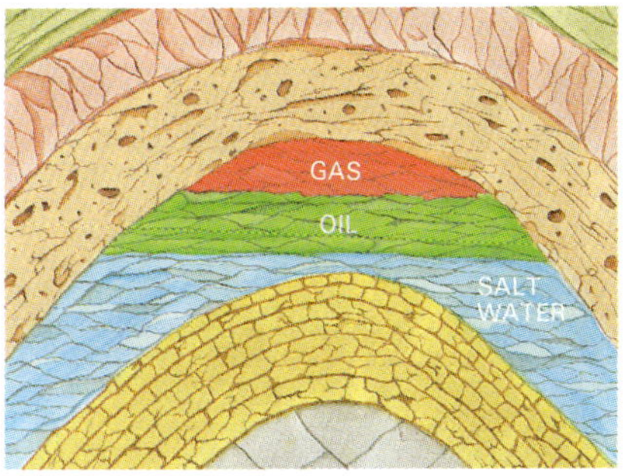

Figure 14. Salt water often lies underneath oil and gas in a reservoir.

Figure 15. Salt water usually coats the grains of the reservoir rock.

The water that is in the pore spaces with oil and gas is called *connate water.* Connate water can even fill up the smaller pores of a rock or coat the rock grains that surround the larger ones. Thus, frequently, a petroleum reservoir not only has oil and gas but salt water as well. This salt water is usually in two places in the reservoir: (1) under the oil and gas and (2) in the pore spaces with the oil and gas.

Oil, gas, and water in a reservoir generally do not occur in a haphazard manner. Instead, gas is in the highest part, oil is below the gas, and salt water is below the oil. (However, remember that salt water, or connate water, also coats the pore spaces where the gas and oil are.) The arrangement of gas, oil, and water happens because the three fluids do not weigh the same—that is, they all have different densities.

Gas is lightest—the least dense—so it is on top. Oil is denser than gas but not as dense as water, so it is in the middle. Since water is the densest of the three, it is on bottom. Generally speaking, gas, oil, and water will be arranged in this way if all three are present in the reservoir. Sometimes, however, not all three fluids occur. There may be only oil and water, only gas and water, or—too bad for the oilman—only water.

Exploration and Drilling

Exploration

Let's imagine we're standing in the middle of a big prairie. Except for sky, all we see in any direction is flat, dry land. Underneath this flat ground, there could possibly be a hydrocarbon trap. But, as we slowly scan the landscape, nothing on the surface indicates the presence of such a trap. This is the same problem faced by those who make their living trying to find hydrocarbon traps.

People who explore for oil and gas traps are called *exploration geologists* and *geophysicists.* Their main job is to find subsurface traps that could contain hydrocarbons. One of the sciences that geophysicists use in their search for traps is *seismology.*

Seismology is the study of vibrations in the earth. The vibrations take the form of sound waves. For instance, an earthquake creates sound vibrations that can be studied by seismology. Sound vibrations can also be made and studied on the surface of the earth. Man-made sounds can help geophysicists find traps.

To understand these special sounds and how they are used to find traps, let's look at a simple comparison. Most of us have experienced the sound of our voices coming back to us from the face of a cliff or building. The sound travels through the air, bounces off the cliff or wall, and returns back to us as an echo. The use of seismology to explore for petroleum traps is based on the same principle, but it is somewhat different.

In seismic exploration, geophysicists are looking for traps buried deep beneath the surface. So, the sound has to travel not through air but through rock. Also, the sound has to have the right characteristics to go through thousands of feet of rock and back up to the surface. Further, the seismic work of a geophysicist usually covers many miles of surface area in order to increase the chances of finding a potential reservoir.

Sometimes, an explosive such as dynamite is used to make the sound on the surface. Regardless of what is used to make it, the sound must be very low in frequency in order to penetrate the many layers of rock. Low-frequency sound is sound that is low in pitch or tone, much like the

Figure 16. A truck-mounted plate is pressed very hard against the ground and vibrated to set up low-frequency sound waves.

lowest note on a piano or organ. (In seismic work, however, the frequency must be even lower than the lowest notes of almost any musical instrument.)

In addition to dynamite, certain other devices are used to make the sounds needed in seismic exploration. For instance, a very heavy steel plate, like a piece of armor plate off a battleship, can be dropped flat onto the surface from a truck specially designed for the purpose. Or the sound can be made by pressing a truck-mounted plate hard against the ground and vibrating the plate (fig. 16). Since the vibrations are not explosions, they don't harm the surrounding environment. Offshore, where explosives could harm marine life, air guns are frequently used. The air gun is a chamber filled with compressed air. As the gun is trailed behind a boat, the air is released suddenly and makes a low-frequency popping noise.

Regardless of how the sounds are made, they enter the layers of rock below the surface, and each rock layer reflects some of the sound back, just as a cliff reflects the sound of a voice back. These reflections of sound, or echos, must be heard or detected. Several sensitive detectors, called *geophones* in land exploration and *hydrophones* in offshore exploration, are used to pick up the echos. On land, the geophones are usually strung out in a line or an array behind trucks and are connected together with electric cable. The cable goes to recording equipment that is usually housed in a truck. Offshore, the hydrophones are trailed in a line behind a boat, and the boat carries the recording equipment.

In either case, when the reflected sound comes bouncing back from the rock layers, it reaches the geophones (or hydrophones, as the case may be) on the surface. The geophones convert the sound vibrations to electric current, which is sent along the cable to a tape recorder. The tape holds a record of the echos. In a laboratory, the tapes are played back into computers, and *record sections* (also called *seismic sections*) are made (fig. 17). Expert interpretation of record sections—a sort of cross section of the earth—can reveal what *may* be a trap for petroleum.

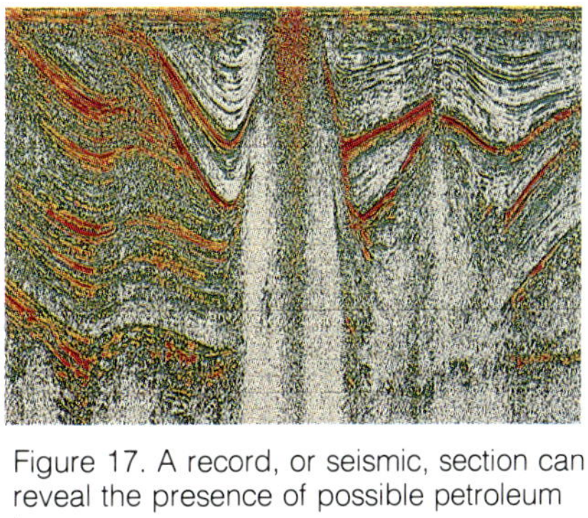

Figure 17. A record, or seismic, section can reveal the presence of possible petroleum traps.

It is important to realize that exploration techniques only indicate where an oil and gas trap *may* be. In most cases, record sections do not give any direct indication of the presence of hydrocarbons in a trap. However, a seismic phenomenon known as a *bright spot* shows up on a seismic section as a sound reflection that is much stronger or weaker than usual (fig. 18). A bright spot sometimes indicates natural gas in a trap. It is safe to say, however, that the only sure way to find out whether hydrocarbons are in a trap is to drill a hole down to it.

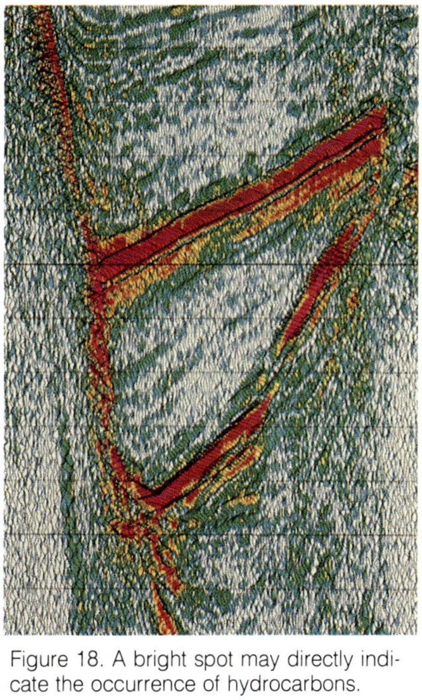

Figure 18. A bright spot may directly indicate the occurrence of hydrocarbons.

Figure 19. A drilling rig is required in order to drill a well.

Figure 20. A drilling bit is made up on the end of the drill pipe and is lowered into the hole.

Drilling

To drill a hole or well requires a drilling rig (fig. 19). Basically, all that happens when a rig drills a hole is that a boring device called a *bit* is pressed hard against the ground and turned. The bit is screwed into the end of hollow pipe called *drill pipe* (fig. 20). A device known as the *rotary table* turns or rotates the pipe and bit connected to it (fig. 21). The rotating bit punches into the rock and scrapes and gouges it out to make a hole (fig. 22).

At the same time the bit is rotating, a special liquid called *drilling fluid,* or *drilling mud,* is pumped down the inside of the hollow drill pipe and out of the bit. Powerful pumps are used to force the mud down the drill pipe (fig. 23). When the mud shoots out

Figure 21. The rotary table turns to rotate the drill string and bit.

Figure 22. The bit teeth penetrate the rock, scraping and gouging away the formation.

Figure 23. A large, powerful pump is required to circulate drilling mud down the drill string and back up the hole.

of the rotating bit, it lifts fragments of rock (cuttings) that the bit has drilled. Then the mud goes up the hole and carries the cuttings with it. At the surface, the cuttings are sifted out of the mud by the *shale shaker* (fig. 24).

Most of the cuttings are thrown away, but the clean mud is pumped back down the drill pipe again. The process of circulating mud down the drill pipe, out of the bit, and back up again is continuous when drilling is going on. The main idea behind circulating the mud is to get the cuttings out of the hole. With the cuttings out of the way, the bit can keep on drilling fresh, undrilled rock.

The hole the rig drills does not simply start at the surface and go all the way down to the hydrocarbon trap in one operation. Instead, the well is drilled in stages. The operation starts when the rig and its parts are moved in and erected over the spot where the hole is to be drilled. If the ground near the surface is soft, a large, round length of hollow pipe is driven or hammered into the ground. If the ground is relatively hard, a hole is drilled with a small rig specially designed for this purpose, and the pipe is lowered into the drilled hole. In either case, the pipe is fairly large—say, at least 20 inches in diameter and 30 or more feet long. The pipe is called *conductor casing.* It is called conductor casing because it conducts the drilling mud from the bottom of the hole back to the surface when drilling starts. Also, the conductor casing keeps the ground from caving in.

To begin the drilling operation, a large bit is screwed onto the drill

Figure 24. Cuttings made by the bit and lifted to the surface in the returning mud flow are separated from the mud at the shale shaker.

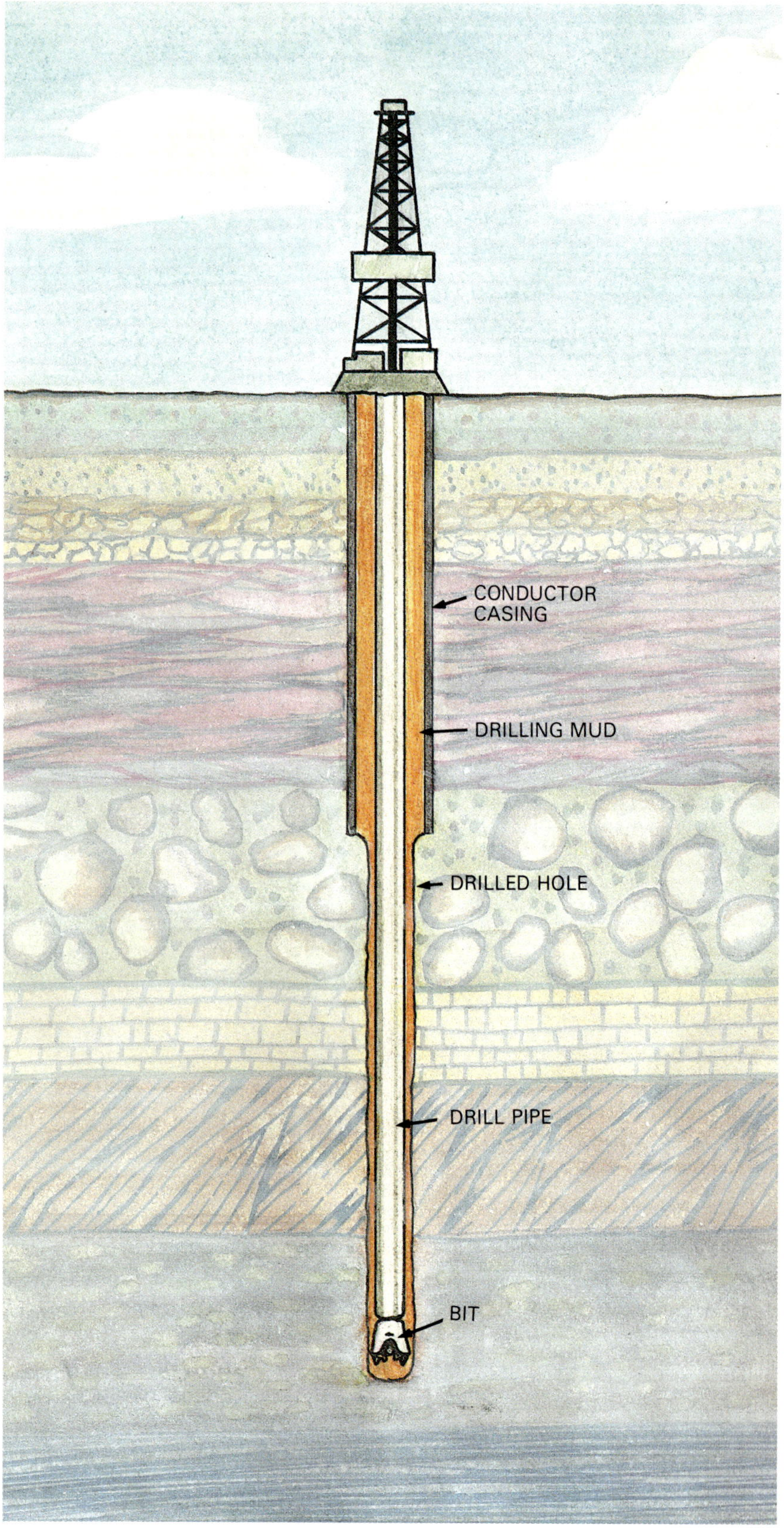

Figure 25. A bit inside the conductor casing drills out the dirt inside the conductor and continues drilling below the conductor.

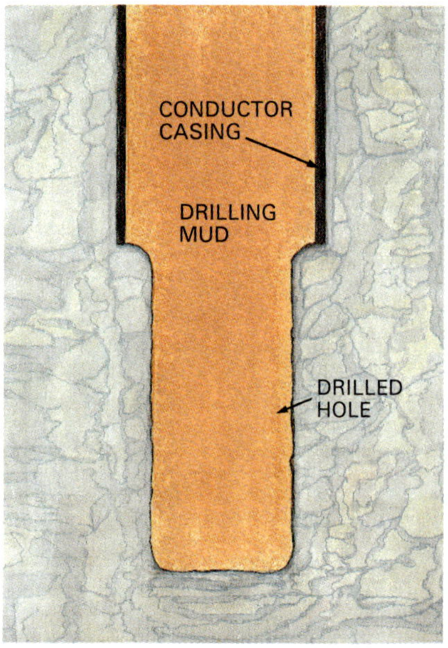

Figure 26. Drilling stops temporarily at a depth well above the trap.

pipe and lowered down to the top of the conductor casing. Assuming that the conductor casing was driven into the ground, the bit is carefully guided into the middle of the opening in the top of the casing. Rotation and circulation are started, and the bit drills out the dirt inside the conductor. Drilling continues past the conductor and into the raw earth below (fig. 25).

At a depth still far above the trap's depth, drilling temporarily stops. The depth of this stopping point is determined before the well is begun and can be hundreds of feet, if not more. All of the drill pipe and bit are then pulled out of the hole. What we have now is a hole in the ground that is full of liquid drilling mud. The top part of the hole is lined with conductor casing (fig. 26).

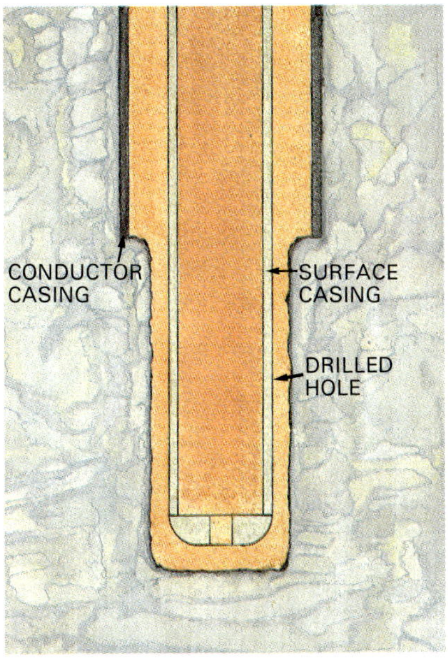

Figure 27. Surface casing is run inside the conductor casing and the drilled hole.

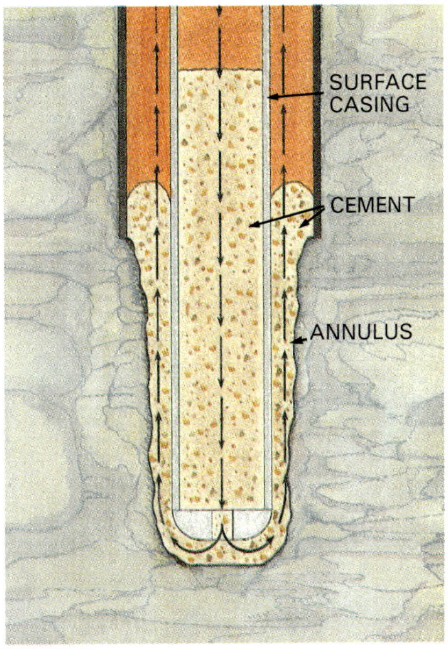

Figure 28. Cement is pumped down inside the surface casing and up the annulus.

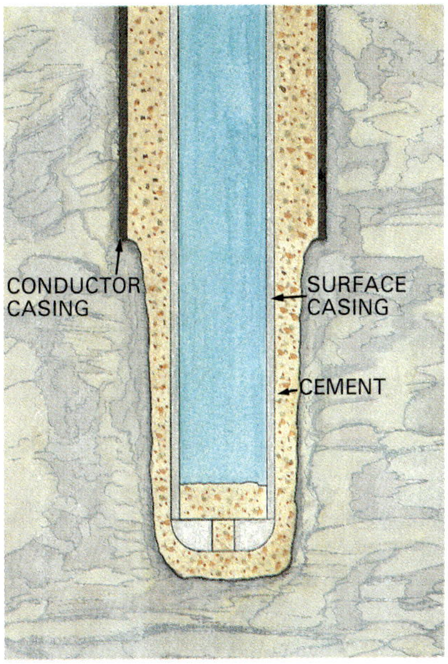

Figure 29. Surface casing is cemented in place inside the conductor and hole.

Why is it necessary to stop drilling the hole before the trap is reached? It so happens that rock layers, or formations, that are fairly close to the surface often contain fresh water. Frequently, towns, cities, farmers, and ranchers use that water for drinking and irrigation. No one wants the water to be contaminated by drilling mud or by oil and gas should the well turn out to be a producer.

Also, formations near the surface, even those below the conductor casing, are often soft and loose. Loose formations can fall into the hole and create problems if something isn't done to prevent it. Therefore, the well drillers drill deep enough to get past the fresh water and the soft and loose formations.

Next, several lengths, or *joints,* of hollow pipe called *surface casing* are screwed together and lowered into the hole. The surface casing goes down inside the conductor casing and inside the hole that was drilled out below the conductor (fig. 27). Now a substance called *cement* comes into the operation.

The cement used in oilwells is very similar to the cement used in ordinary concrete to make bridges, sidewalks, and the like. (Do not make the mistake of calling concrete *cement.* Cement is only one component that goes into the making of concrete.) Cement is pumped down inside the surface casing, out the bottom, and back up the space between the casing and the hole in which the casing hangs. This space is called the *annulus.*

After the cement is pumped, the inside of the casing will not have much cement in it because the cement is pumped until most of it has gone out and around the bottom of the casing and up into the annulus (fig. 28). When the cement sets or hardens, it bonds the casing very tightly to the wall of the hole and keeps water and any other fluids confined in the formation. What we end up with at this point is cemented surface casing hanging inside conductor casing (fig. 29).

The next step is to use a bit that will fit inside the surface casing and get back to drilling. If the trap does not lie too much deeper—say, a couple of thousand feet or so—the part of the hole that is drilled out below the surface casing could be the part that goes on down to the trap. On the other hand, if the trap is really deep—say, 10,000 feet or more—the part of the hole below the surface casing will be just another step in the process. In other words, drilling will stop once more and another string of casing will be run and cemented, even though the trap has not been reached. Whenever it is necessary to drill a deep well, the chances are good that the hole will go through so-called troublesome formations. Troublesome formations are formations that can be successfully drilled but have to be lined with casing and cement if the well is ever going to be successfully produced.

For example, one kind of troublesome formation is one with high-pressure fluids in it. A formation with high pressure can blow out through the hole and into the air if care is not taken when drilling it. Blowouts can be impressive and dangerous (fig. 30). The people who drill wells usually handle high-pressure zones with few problems, but occasionally a well will get out of control and blow out. In any case, after a high-pressure formation is drilled, it is sealed off with casing and cement. Casing is made of steel and is designed to withstand high pressure.

When casing is needed below the surface casing, it is called *intermediate casing.* Intermediate casing is hung inside surface casing and runs from the surface all the way to the bottom of the hole. Intermediate casing, like surface casing, is cemented in the well (fig. 31). As stated before, every well does not require intermediate casing. Only wells that run into troublesome formations below the surface casing need it. The deeper the hole, the better the chance of encountering troublesome formations. In fact, some very deep wells may require more than one string of intermediate casing.

Once intermediate casing or, in some instances, surface casing is run and cemented, the next step is to drill more hole—this time down to the trap that presumably has hydrocarbons in it. When the borehole reaches the trap, the well owners have to decide whether it contains enough oil and gas to make it worth the money to finish up, or *complete,* the well.

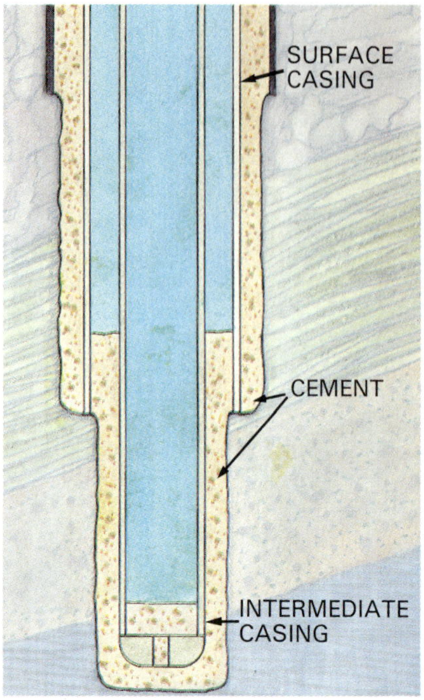

Figure 31. Intermediate casing is cemented in the well.

Figure 30. A blowout and fire reduced this rig to molten steel.

Testing and Completion

It costs a lot of money to drill a well. It costs still more to complete one. If the well owners are going to spend the money, then they need to know whether there are enough hydrocarbons in the trap to justify the expense of completion. The figure varies from year to year, but on the average, for every seven exploratory, or *wildcat,* wells drilled, six of them will be dry. In other words, in the long run only about one wildcat well in seven actually strikes oil or gas. Pretty bad odds, but if one really good reservoir is found, it will make up for the bad ones—at least that's the hope. In any case, once the hole goes into the reservoir, or what everyone hopes is a reservoir, the well owners have to run some tests to see if hydrocarbons are present.

Formation Testing

One way to test a well is to *log* it. *Well logging* involves lowering a special instrument into the well. As the logging instrument is pulled back up the hole, it sends signals to the surface where electrical, radioactive, or acoustic properties of the formations are recorded. The record or log has several lines or curves on it (fig. 32). Experts look at the curves and can pretty well tell whether a formation has petroleum in it and how much may be there.

Another way in which to test a well is to take *core samples.* A special coring drill is used to obtain the samples. A core is simply a long cylinder of rock, small in diameter, that is obtained from a formation of interest. When brought to the surface, the core is analyzed and can reveal the porosity and permeability of the rock, the presence of oil or gas, and how much may be there.

A *drill stem test* is another method of testing a formation. The drill pipe and other pipe components used in drilling a well are called

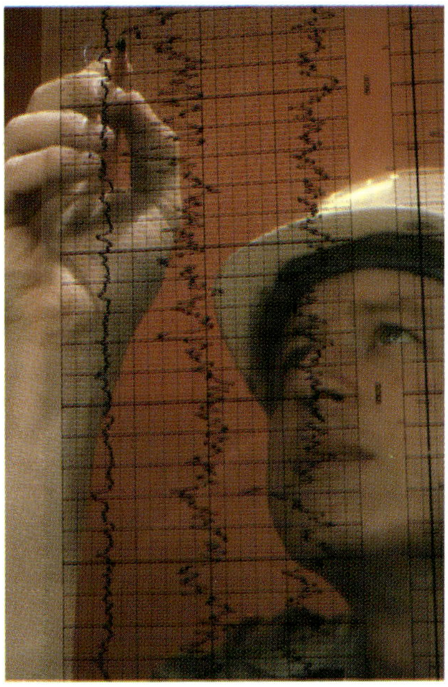

Figure 32. A well log is examined to see whether it indicates the presence of hydrocarbons and other fluids in a formation.

the drill stem. When a special testing tool is attached to the bottom of it, a drill stem test can be run. The tool is lowered into the hole on the drill stem until it reaches the formation to be tested. The tool is activated and the test run. Basically, the test allows the well to flow for a period of time and measures and records the pressure in the formation. From the fluids produced and the pressure record, information about the formation can be deduced. The kinds of fluids in the formation, the flow rate of these fluids, the permeability of the formation, and the pressure in the reservoir are a few of the things that a drill stem test will tell an expert.

Usually, one or more of the tests will indicate whether or not the trap has enough hydrocarbons to make it worthwhile to complete the well for production. If the well tests indicate that the hole is a *duster*—a dry hole that has little if any hydrocarbons—the hole will be *plugged and abandoned.* Several hard cement plugs will be put into the well at several different depths and at the surface. The plugs permanently seal up the well. However, if test indications are favorable, and it appears that the trap contains an ample amount of hydrocarbons, then one more string of casing is run and cemented in the well. This final casing string is the *production casing.* It is termed *production casing* because it passes through the producing formation.

Perforated Completions

For purposes of illustration, let's assume that we have a possible producer. It has conductor, surface, intermediate, and production casing in it. All the casing strings are cemented except the conductor, and sometimes it is cemented

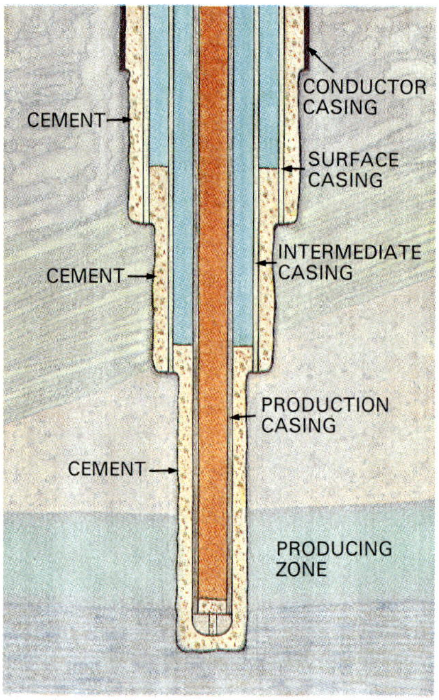

Figure 33. Conductor, surface, intermediate, and production casing are cemented in the well. Note that the production casing is set through the producing zone.

(fig. 33). As mentioned earlier, not all wells have intermediate casing, and some have more than one string of intermediate casing. Looking closely at figure 33 reveals that production casing has been cemented through the producing formation or zone. At present, the hydrocarbons cannot get out of the formation and into the production casing because cement and casing seal it off. Obviously, something further must be done. The well must be *completed.* One technique used for well completion is *perforating.*

For a *perforated completion,* a special tool—a perforating gun—is lowered down the production casing until it is opposite the producing zone. Then, the gun is fired to shoot *bullets* or to set off special explosive charges known as *shaped charges.* A shaped charge is designed to go in one direction, so that an intense, directional explosion is formed. The explosion is actually a jet of high energy gases and particles. Usually, several charges are loaded into the gun. When set off, the jets or bullets form holes or perforations in the casing and cement and force their way a short distance into the formation. The perforations provide openings for the hydrocarbons and other fluids (fig. 34). When a perforating gun is used, the technique is called a perforated completion. A perforated completion is the most common, but several other methods are also used to complete a well.

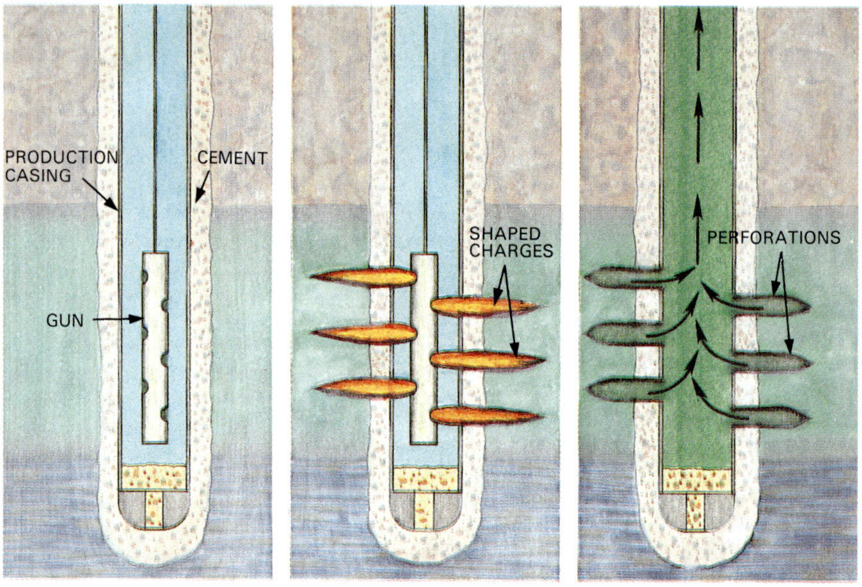

Figure 34. A perforating gun is used to create holes or perforations in the casing, cement, and producing formation, allowing petroleum in the reservoir to enter the well.

Open-Hole Completions

In a few instances production casing is not cemented through the producing zone. Instead, it is set and cemented above the reservoir. The zone itself is left open to the uncased hole drilled into it. This is termed an *open-hole completion* (fig. 35). Reservoir fluids simply flow into the open, uncased hole. Open-hole completions are very rare because the formation must be composed of hard, homogeneous rock, and not too many reservoirs fit that description.

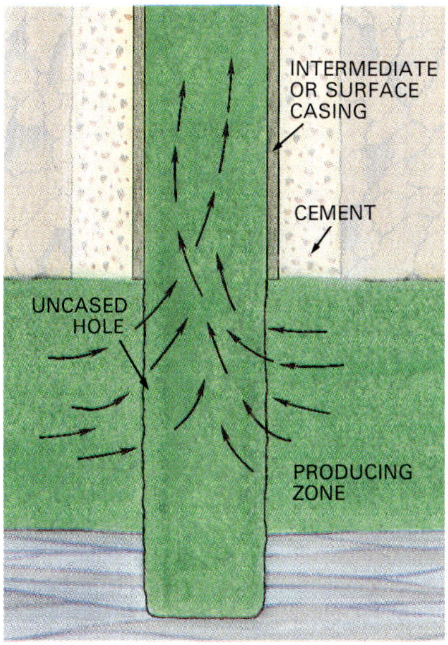

Figure 35. An open-hole completion allows reservoir fluids to flow into the uncased hole.

Perforated Liner Completions

Another completion method similar to a perforated completion is a perforated liner completion. What's a liner? Recall what production casing is: round, steel pipe that runs from the surface down through the producing zone. Suppose, however, that instead of running all the way back to the surface, the top of the casing is hung inside and near the bottom of the intermediate or surface casing. Such a shortened length of casing is a *liner* (fig. 36). In other words, a liner is casing that does not reach back to the surface. A production liner can be cemented through the producing zone and perforated. This is a perforated liner completion (fig. 37).

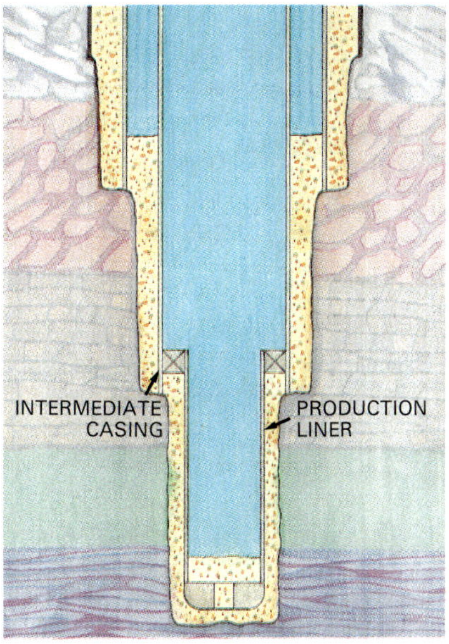

Figure 36. A production liner is hung inside a previously run string of casing.

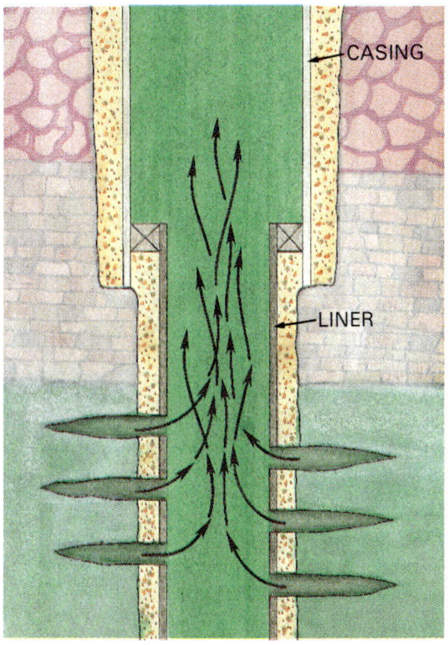

Figure 37. In a perforated liner completion, the shortened length of casing, or liner, is perforated opposite the producing zone.

Wire-Wrapped Screen Completions

Another type of completion involves a tubular device called a *wire-wrapped screen*. Basically, a wire-wrapped screen is a relatively short length of pipe that has openings in its sides. A specially shaped wire is wrapped around the pipe. This screen is attached to the bottom of a *tubing string* (tubing is simply small-diameter pipe placed inside the casing) and is lowered into a well that has been perforated. Usually, wire-wrapped screens are run in conjunction with a *gravel pack*. The gravel, which actually resembles coarse sand, is placed in the hole outside the screen. Formation fluids flow through the gravel pack, through the wire-wrapped screen, and into the tubing (fig. 38).

Generally, wire-wrapped screen completions with gravel packs are employed in producing zones that consist of loose sand. Loose sand can flow with the produced fluids into the well and create problems. For instance, sand can clog up the flow and reduce or stop production. It can also erode equipment. The screen, in combination with the gravel pack, blocks out the sand and allows fluids to flow into the well through the openings in the wire-wrapped screen.

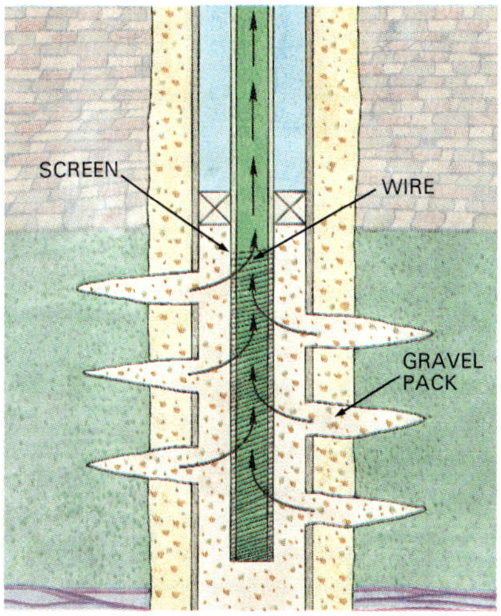

Figure 38. A wire-wrapped screen completion is often run with a gravel pack inside perforated casing.

Tubing and Packers

Regardless of how a well is arranged to allow fluids to flow into it, most well completions entail a further step. One more string of pipe is run. This pipe, called *tubing,* is relatively small in diameter and extends from the surface to the bottom of the well (fig. 39). Petroleum and other fluids from the reservoir actually flow to the surface through the tubing string. Why aren't wells produced through casing?

Casing or liners will remain in the well throughout its producing life. If reservoir fluids are corrosive to steel, or if anything else goes wrong with the casing, it will have to be removed and repaired—a difficult and very costly operation. Also, the diameter of the casing is usually too large to allow the well to flow efficiently. So, to avoid casing problems and to allow the well to produce better, tubing is used as a flow string. Should problems occur with the tubing, it can be pulled and replaced.

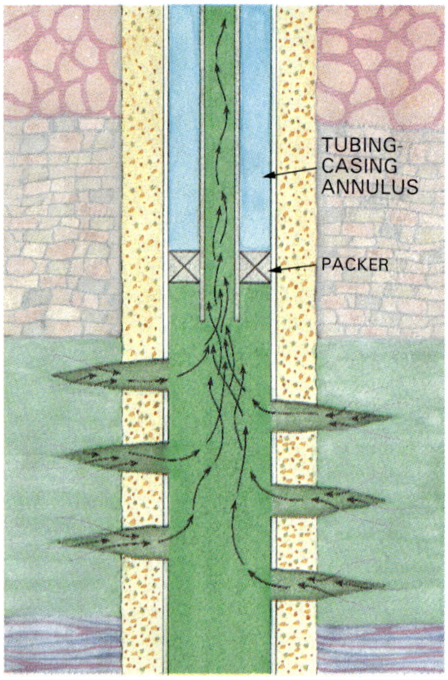

Figure 40. A packer placed in the tubing string keeps reservoir fluids out of the tubing-casing annulus.

Often, a device called a *packer* is placed in the tubing string. A packer seals off the annulus between the tubing and production casing at a point just above the producing zone. The packer is attached to or, as they say in the oil patch, *made up* in the tubing string. It consists of pipe through which well fluids can flow. It has gripping elements called *slips* that grip the wall of the casing. The slips anchor the packer. A packer also has sealing elements that are usually made of rubber. When expanded, the sealing elements provide a pressure-tight seal in the casing-tubing annulus. Thus, when formation fluids flow into the well, the packer forces them to flow into the tubing (fig. 40).

Another device frequently installed in the tubing string is a *subsurface safety valve.* It is made up on the tubing string fairly close to the surface. As long as fluid flow in the tubing is normal, the valve stays open and fluids continue to flow. However, if some upset happens—say, a truck runs into the well surface equipment and well fluids start flowing into the air or on the ground—the safety valve senses the loss of control and closes. The closed valve prevents the further escape of fluids (fig. 41).

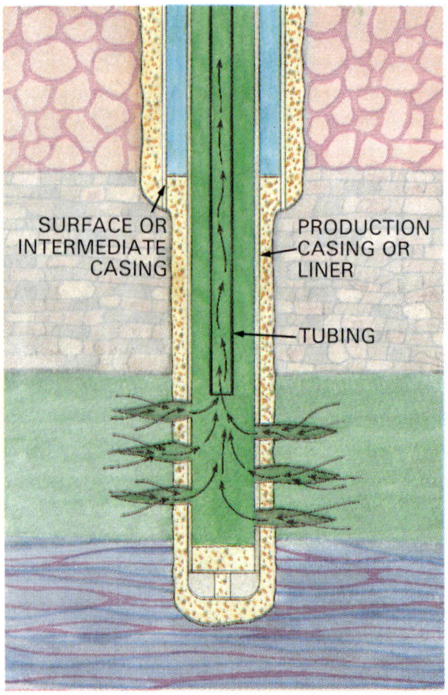

Figure 39. Tubing is run in a perforated well so that reservoir fluids can flow to the surface efficiently.

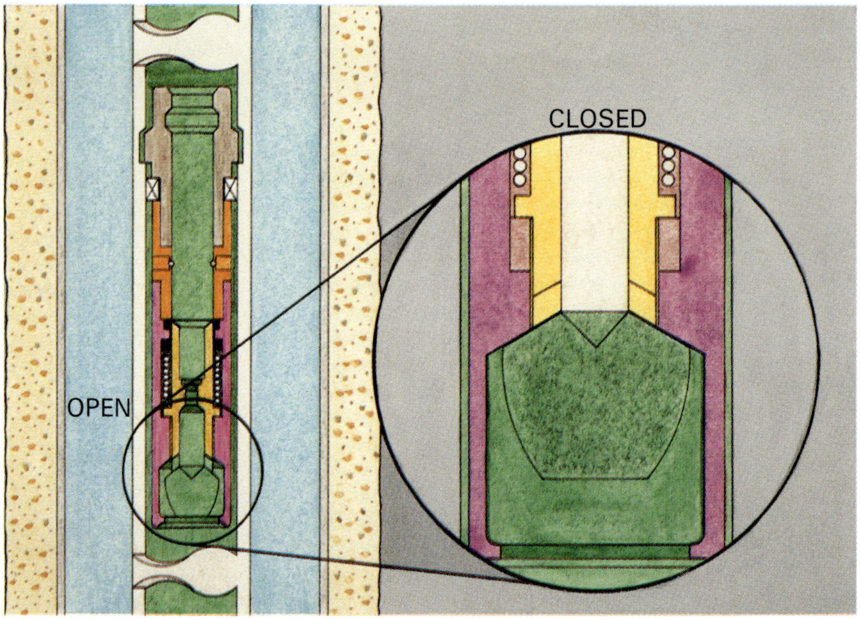

Figure 41. A subsurface safety valve is installed in the tubing string. When the valve is open (*left*), well fluids flow through the valve. When the valve is closed (*right*), well fluids are contained in the tubing.

Tubingless Completions

While most wells are completed with tubing, some are not. In some instances, it is possible to drill a small-diameter well and use small-diameter casing to conduct reservoir fluids to the surface. In such cases, the well is said to be a *tubingless completion* (fig. 42). The small casing is cemented and perforated opposite the producing zone. Except for the lack of an inner tubing string, a tubingless completion is fairly conventional. Tubingless completions are used mostly in small gas reservoirs that produce few if any liquids and are low in pressure.

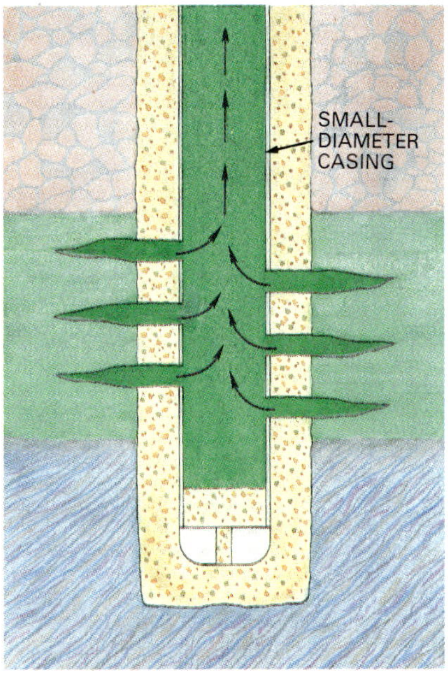

Figure 42. A tubingless completion has no inner tubing string. Well fluids are produced through small-diameter casing.

Multiple Completions

Another special kind of completion is a *multiple completion.* Multiple completions are used when one wellbore penetrates two or more producing zones. Usually, a separate tubing string and packer for each of the zones is run inside a single string of production casing. For instance, in a triple completion, three tubing strings and three packers can be utilized (fig. 43).

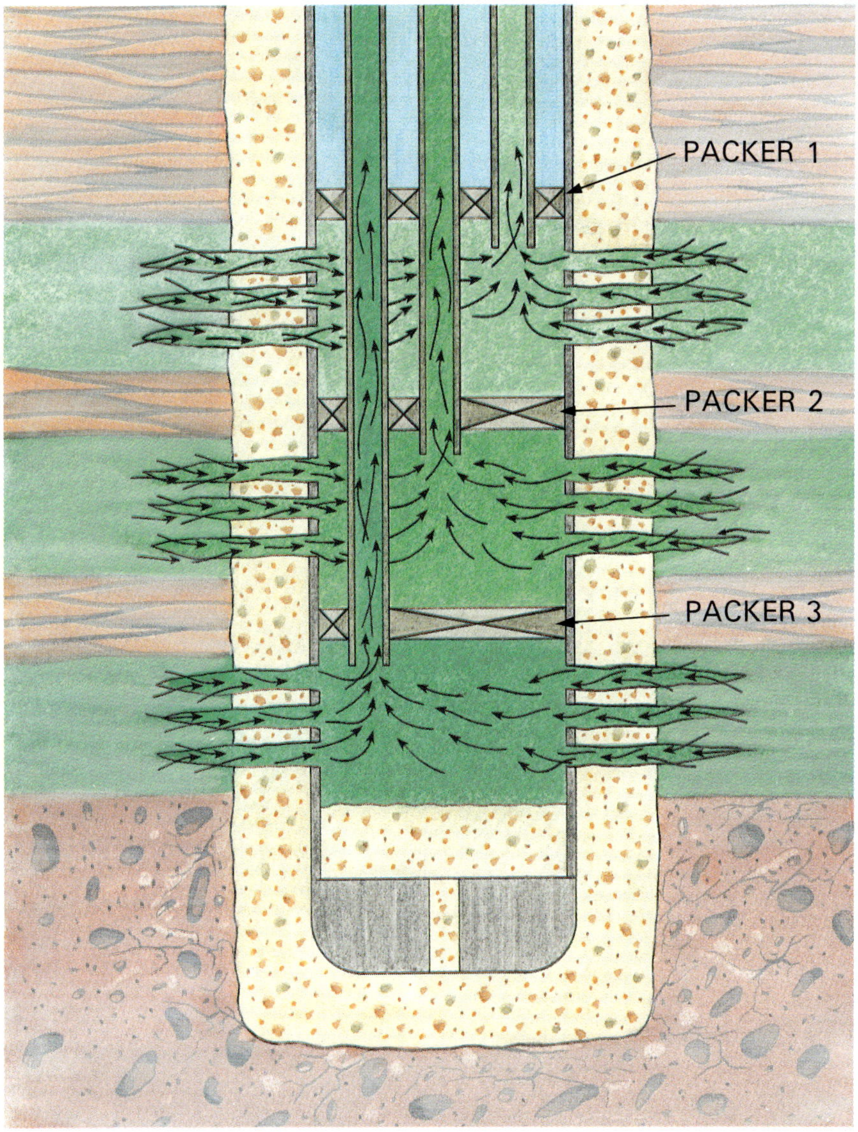

Figure 43. A multiple completion, such as this triple, usually has a separate tubing string and packer for each producing zone. Note that each zone is isolated from the others.

Reservoir Stimulation

Petroleum and other reservoir fluids like water are found in the pore spaces of the rock making up the reservoir. These pore spaces must be connected so that the fluids can flow out of the rock and into a well. Sometimes, a reservoir rock has very little permeability. A reservoir with low natural permeability can present a problem because the hydrocarbons in the reservoir, though ample, cannot be extracted at reasonable rates.

Formation damage can also prevent hydrocarbons from being extracted at reasonable rates. Formation damage is a man-made problem. The natural permeability of the reservoir rock may be adequate, but when a well is drilled into the reservoir, the permeability of the rock formation near the reservoir is reduced.

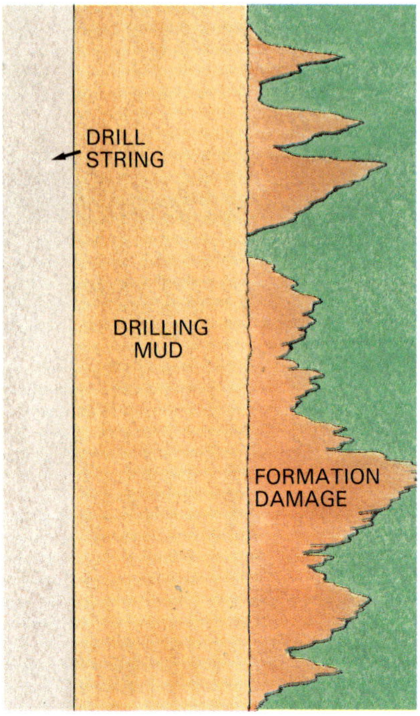

Figure 44. When the hole is being drilled, drilling mud can seep into a permeable formation and damage or block the permeability.

Formation damage is caused by many things, but one common cause is the drilling mud being used at the time the hole is drilled into the reservoir. The mud comes into contact with the reservoir formation and can cause damage. The mud can actually seep into the formation and block the permeability (fig. 44). In other cases, the water in the mud can block or reduce permeability of the formation. As a result, hydrocarbons cannot readily flow into the well.

Techniques are available to remove or reduce formation damage and to overcome the problem of low natural permeability. The two most popular reservoir stimulation techniques are *fracturing* and *acidizing.*

Fracturing

One way in which to stimulate a reservoir is to break, or fracture, it. Fracturing is done after the well has been completed—that is, the well has casing, tubing, and perforations.

Fracturing actually splits open the rock reservoir. The technique uses several powerful pumps arranged on the surface near the well to be fractured (fig. 45). The pumps are connected by high-pressure lines to the well tubing. A special fluid mixture, often composed mainly of water and sand, is pumped down the tubing to the bottom of the well. Pumping continues, and pressure forces the fracturing fluid into the perforations. Soon, the pressure becomes so high that it overcomes the strength of the reservoir rock and cracks it open. Pumping continues until the fracture is of the desired length. Then pumping ceases, the pressure is bled off, and the fracturing equipment is removed. The sand in the fracturing fluid holds the fracture apart so that reservoir fluids can flow into the fracture and to the well (fig. 46).

Fracturing a reservoir is similar to splitting a log with a hammer and wedge. Muscle power swings the hammer to drive the wedge into the log to split it open. In a fracturing job (often shortened to *frac job*), powerful engines replace

Figure 45. Several powerful, truck-mounted pumps are arranged at the well site for a fracturing job.

muscles, and high-pressure pumps take the place of the hammer. Fracturing fluid becomes the wedge, and the formation behaves the same as the log—it splits open.

Now picture the split log being held open by the wedge (the log is not split in two). What happens if the wedge is taken out? The split comes back together, or heals. A fracture in a reservoir behaves in the same way. When the pumps are stopped and pressure bled off, the fracture comes back together—or will if allowed to. However, the sand in the fracturing fluid holds, or props, the fracture open, preventing it from healing. Because sand serves this function, it is often called a *proppant.* The propped fracture serves as a pathway for reservoir fluids to flow into the well and to the surface.

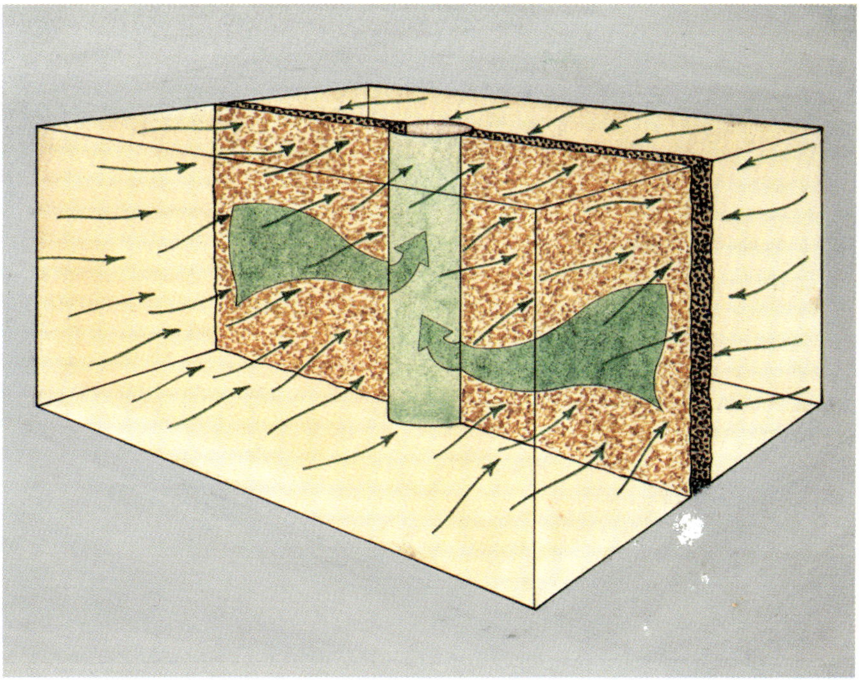

Figure 46. Reservoir fluids flow into the propped fracture and well.

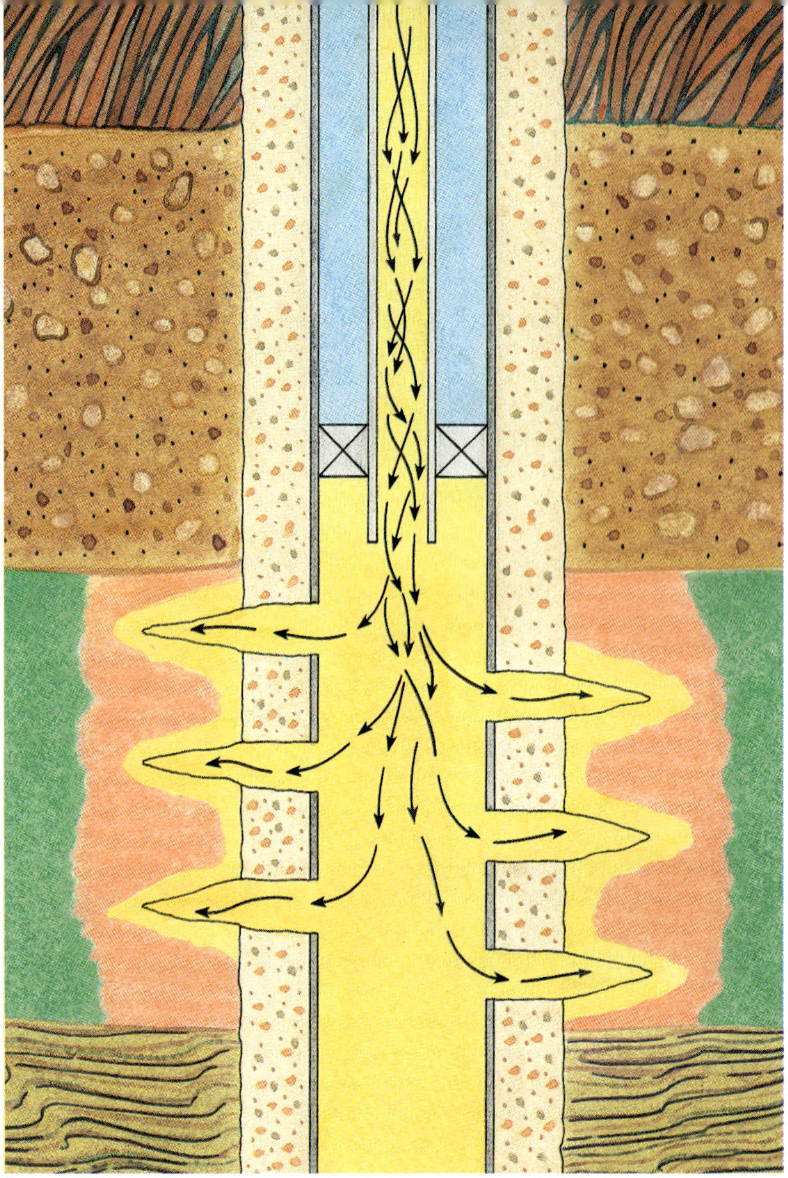

Figure 47. Acid is injected down the tubing and into the formation through perforations to remove formation damage without fracturing the formation.

Acidizing

Certain acids, when put into contact with suitable reservoir rocks, etch or dissolve them. The fact that acid reacts with rock forms the basis of a stimulation method known as *acidizing.* Two types of oilwell acidizing are fracture acidizing and matrix acidizing.

Fracture acidizing involves pumping acid into the reservoir at a pressure high enough to cause it to fracture. Fracture acidizing is similar to fracturing; however, acid instead of water is used as a fracturing fluid. The acid etches the face (each side) of the fracture to create flow channels for reservoir fluids. Fracture acidizing is utilized mainly in limestone or similar formations.

Matrix acidizing does not involve fracturing. (The body of the rock in which hydrocarbons and other fluids occur is often known as the rock matrix.) In this method acid is injected down the tubing and into the permeable channels of the reservoir. The injection pressure is kept low enough to prevent fracturing the formation. Certain acids can remove substances that block reservoir permeability, especially mud solids (fig. 47).

Many different kinds of acids are used in acidizing oilwells. One of the more common ones is *hydrochloric acid.* Hydrochloric acid is a strong acid that works best on carbonate rocks. Carbonate rocks include limestone and dolomite. Limestone is mostly calcium carbonate, and dolomite is mostly calcium magnesium carbonate. Limestone and dolomite reservoirs are called *carbonate reservoirs.*

Acetic acid and *formic acid* are sometimes used to acidize carbonate reservoirs that have high temperatures—250°F or more. Another acid used is *hydrofluoric,* or *mud, acid.* Hydrofluoric acid reacts with quartz, sand, and clay. Also, various acids can be mixed in order to obtain the desired results in an acidizing job. Regardless of the acid used, the idea behind acidizing is to enable reservoir fluids to flow out of the reservoir and into the wellbore.

Reservoir Drives

After a well is drilled and completed (and if necessary, stimulated), attention turns to producing oil and gas from the reservoir. Some reservoirs contain mostly gas and others mostly oil. Let's consider those that contain mainly oil first.

In order for oil to flow to the surface, there must be a drive force associated with it in the reservoir. Fortunately, at the time oil was forming and accumulating in reservoirs, *pressure energy* in the gas and salt water associated with the oil was also being stored. This gas or water pressure (or both) is what drives oil through and from the pores of the reservoir, into a well, and to the surface.

Dissolved-Gas Drive

In most cases, oil in a reservoir has gas dissolved in it. This gas comes out of the oil when pressure drops as the oil is withdrawn through wells drilled into the reservoir. As the gas escapes from the oil, it expands. The expanding gas drives oil through the reservoir toward the wells and assists in lifting it to the surface. Reservoirs in which oil is lifted by dissolved gas escaping and expanding from the oil are termed *dissolved-gas drive reservoirs* (fig. 48).

Unfortunately, dissolved-gas drive is not very effective in terms of the amount of oil it allows to be recovered. Typically, what happens is that oil flows to the surface only at the beginning. Rather quickly, reservoir pressure drops to a value where it can push oil only to the well and not to the surface. At this time, some type of pump or other *artificial lift method* must be applied to recover more oil. However, a point will eventually be reached where no more oil can be recovered, even with artificial lift. In fact, statistics show that only 5 to 30 percent of the oil originally in place in a dissolved-gas drive reservoir can be recovered. Any portions of the remaining 70 to 95 percent can be produced only by additional recovery techniques.

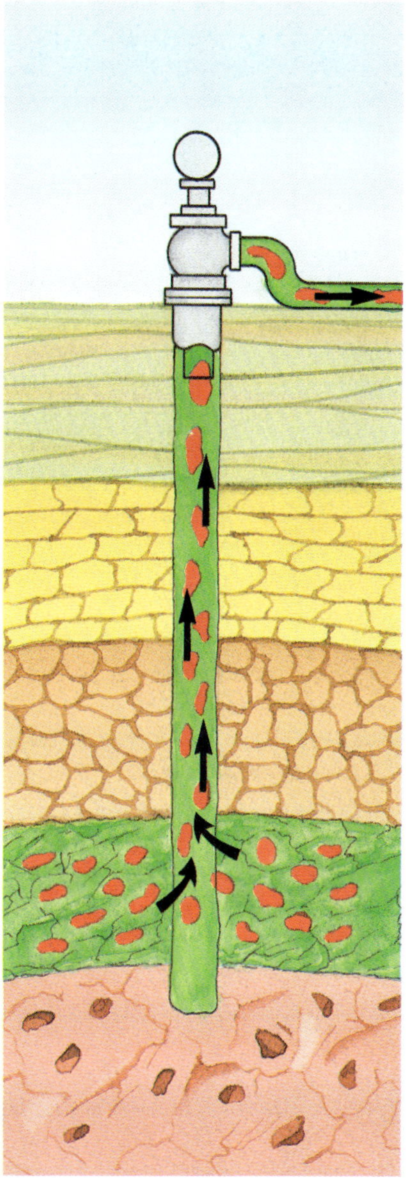

Figure 48. In a dissolved-gas drive reservoir, gas comes out of the oil, expands, and lifts oil to the surface.

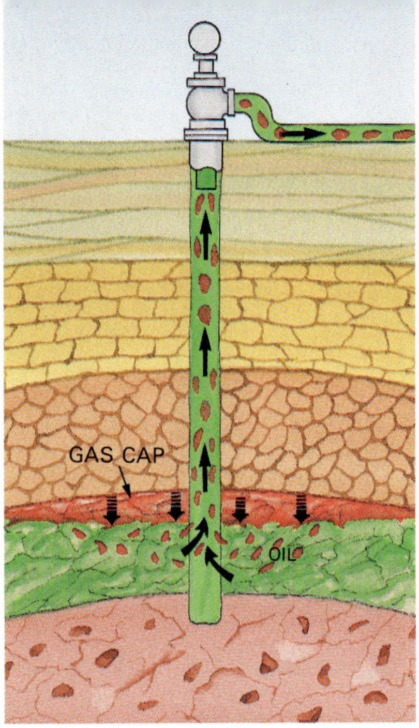

Figure 49. In a gas-cap drive reservoir, the free gas in the cap and the gas dissolved in the oil expand to move oil up to the surface.

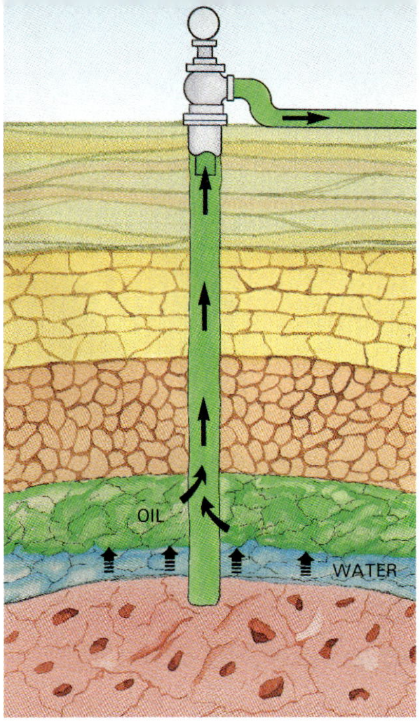

Figure 50. In a water drive reservoir, salt water under the oil invades to lift oil to the surface.

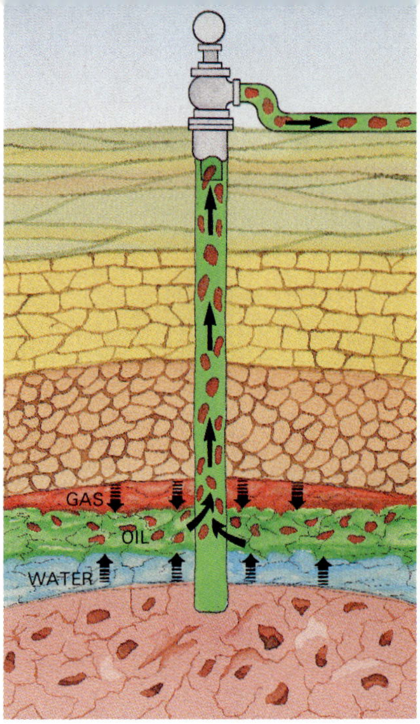

Figure 51. In a combination drive reservoir, expanding gas in the cap, gas dissolved in the oil, and water under the oil act to move oil to the surface.

Gas-Cap Drive

Some oil reservoirs have so much gas that all of it cannot be dissolved in the oil. This extra gas forms as a layer, or *cap,* over the oil (fig. 49). In gas-cap drive reservoirs, two gas drives are actually at work. One is the gas dissolved in the oil; the other is the gas in the cap. As oil is produced through wells drilled into the reservoir, the dissolved gas comes out of the oil, expands, and helps drive oil to the surface. In addition, the gas cap expands to push oil to the surface.

Gas-cap drive is a more effective drive than dissolved-gas drive alone. A reservoir with a gas cap may yield from 20 to 50 percent of the oil originally in it.

Water Drive

Let's visualize a porous and permeable rock layer buried deeply beneath the surface, overlain by an impermeable layer, and covering a very large area. In the highest part of the porous layer lies an oil deposit that is relatively small in size compared to the rest of the layer. Occupying the remainder of the pores of the layer, and lying in contact with the oil, resides a vast amount of salt water (fig. 50). This huge quantity of salt water occurs under pressure and provides a source of energy for driving oil to the surface.

When wells are drilled into the oil zone of a formation that contains large amounts of water adjoining the oil, the water moves in behind the oil as it is produced. The invading water pushes the oil out of the reservoir, into the wells, and to the surface. Such a drive is called *water drive* and is the most effective drive of all. In fact, from 35 to 75 percent of the oil in the reservoir can be produced by water drive.

Combination Drive

A combination drive, a drive in which both free gas in a cap and water are available in the reservoir, is the most common type of drive (fig. 51). In combination drives, the gas and water expand and displace oil to the surface. In addition to the free gas in the cap, gas dissolved in the oil, which escapes and expands when wells are produced, is a factor in combination drives.

In reservoirs that contain hydrocarbons primarily in the form of free gas, the drive force can be thought of as a combination drive. Even though the hydrocarbons in the reservoir are mostly gas, frequently hydrocarbon liquids (often called *gas liquids*) and water are associated with them. While wells drilled into such reservoirs are termed *gas wells,* the expanding gas and water will also drive any liquid hydrocarbons as well. Of course, in reservoirs where no massive amounts of water are associated with the gas and liquid hydrocarbons, then gas-cap drive would be in force.

Wellheads

We've discussed some of the equipment, techniques, and processes that are used and occur in the subsurface. Let's turn our attention to some of the equipment that is installed on top of a well.

The *wellhead* is the equipment used to confine and control the flow of fluids from the well. Wellheads vary in size, strength, and configuration, depending on well conditions. For example, a wellhead could consist of several heavy fittings, with certain parts designed to hold very high pressures. Such a wellhead would be used on a high-pressure gas well. On the other hand, it could be just a simple assembly designed mainly to support tubing in the well—say, in cases where pressure and production are low. However, it is possible to say that a typical wellhead is made up of a *casinghead, tubing head,* and *Christmas tree* (fig. 52). Bear in mind, though, that there is a lot of variation in wellhead design and installation.

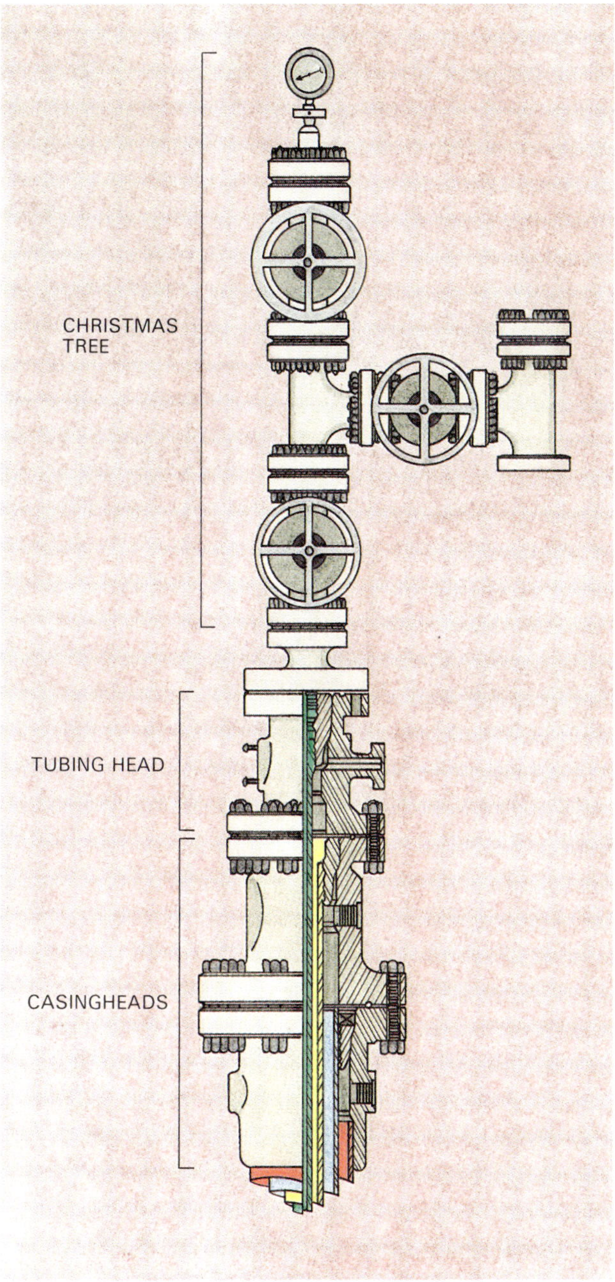

Figure 52. A typical wellhead is made up of a casinghead (sometimes more than one), a tubing head, and a Christmas tree.

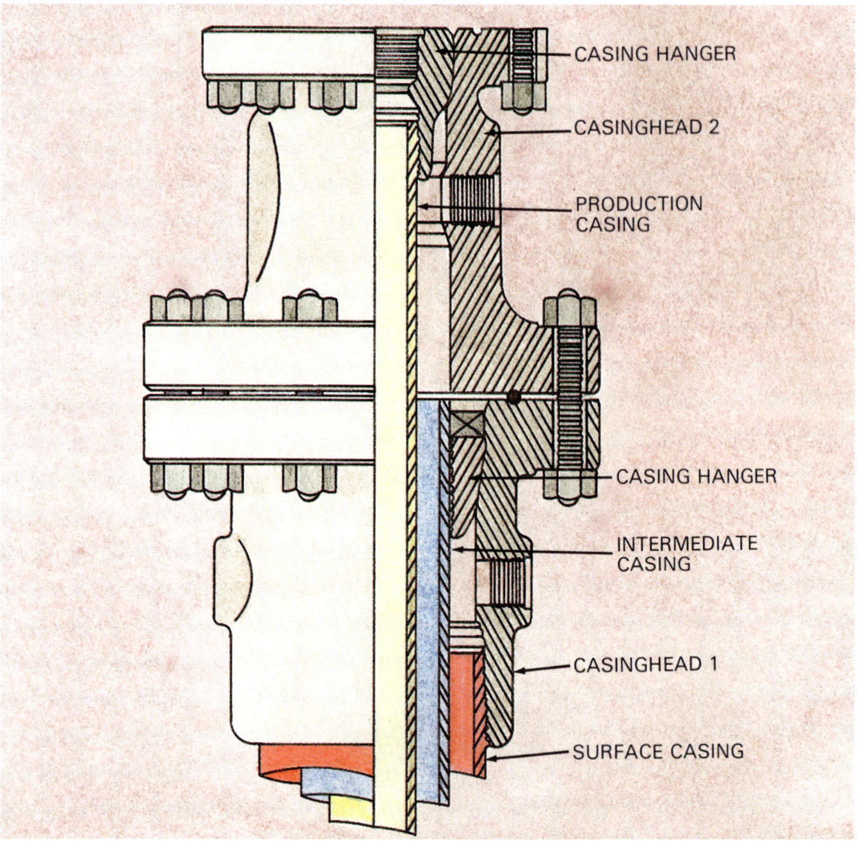

Figure 53. Two casingheads are used on top of three strings of casing in this example. Note that the first casinghead is screwed onto the surface casing. The other casing strings are suspended in the heads with hangers.

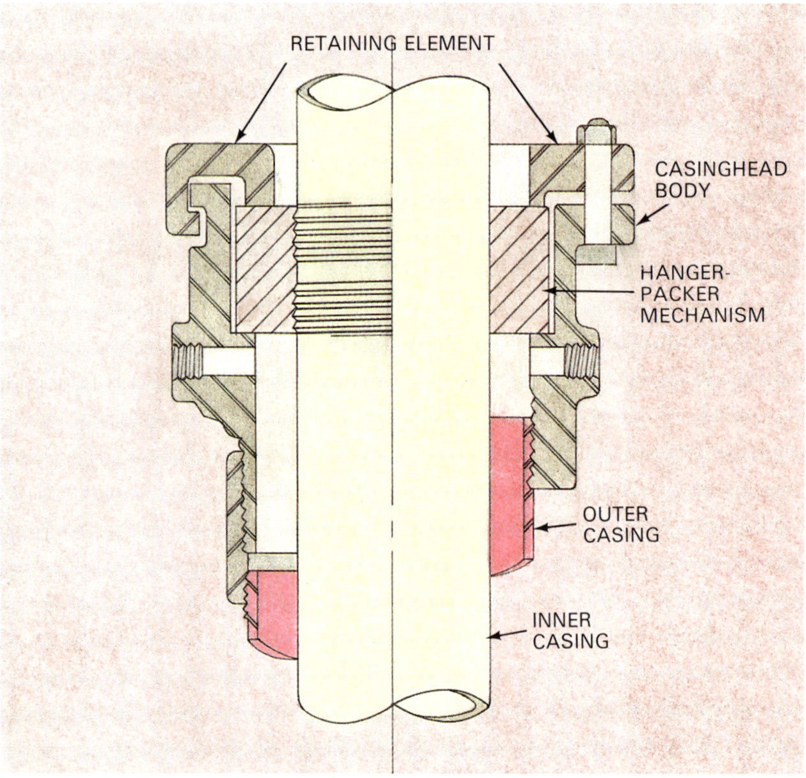

Figure 54. Less complex casingheads may be used in wells that have only two casing strings.

Casinghead

A casinghead consists of several parts attached to the top of each string of casing after each string has been run and cemented in the well. In wells that have an intermediate casing string, two casingheads may be used, one on top of the other (fig. 53). In wells having only surface (outer) casing and production (inner) casing, the casinghead may be less complex (fig. 54).

Regardless, the casinghead provides for the use of gripping devices, or *hangers,* to help support the weight of the casing. Casingheads also have sealing elements or packers to prevent the flow of fluids within the head and to the atmosphere. Openings are provided for bleeding off pressure that may collect in or between casing strings.

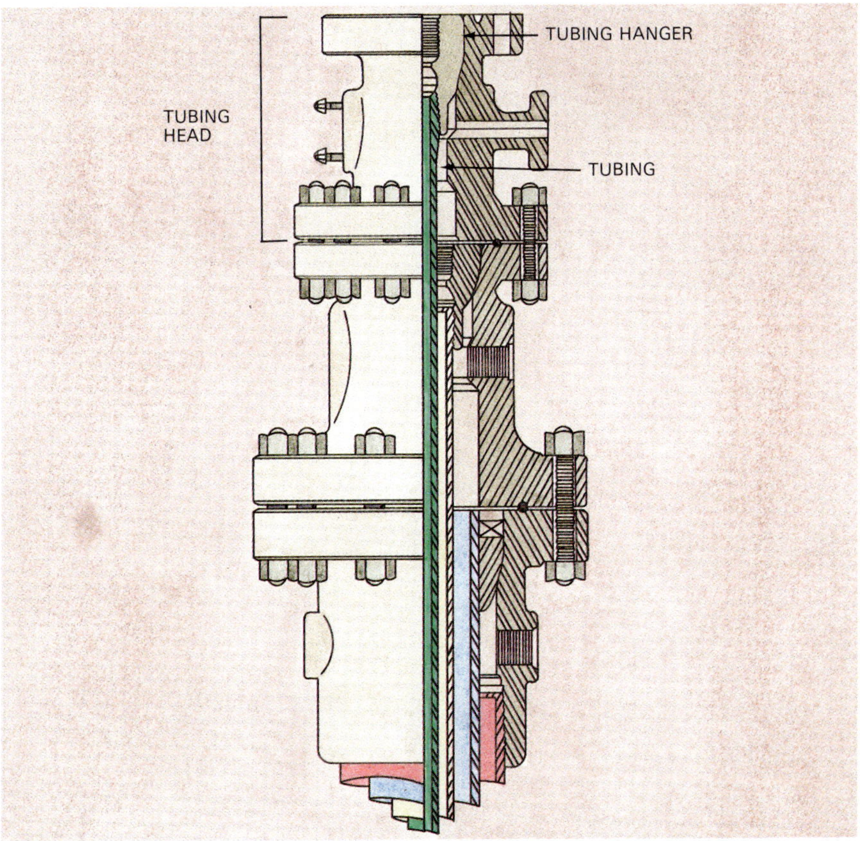

Figure 55. A tubing head can be installed on top of the casinghead. The tubing string is hung from the tubing head. Note that all the individual strings of casing and the tubing are isolated from each other so that no pressure or fluids in one string can get into another.

Tubing Head

Similar in design and use to a casinghead, a tubing head is primarily used to support the tubing string. It also seals off pressure in the casing-tubing annulus. In addition, it provides connections whereby fittings and valves can be attached to control the flow of fluids. The tubing head is made up on the casinghead, where casingheads are used (fig. 55).

In low-pressure wells, a casinghead may not be used. In such cases, the tubing head is connected directly to the top of the casing (fig. 56). In addition to outlets that provide a way to reduce any gas pressure in the casing, such tubing heads may have a *hanger-packer mechanism*. The hanger-packer not only supports the tubing, but seals the outside of it as well. As with casingheads, tubing heads vary in design.

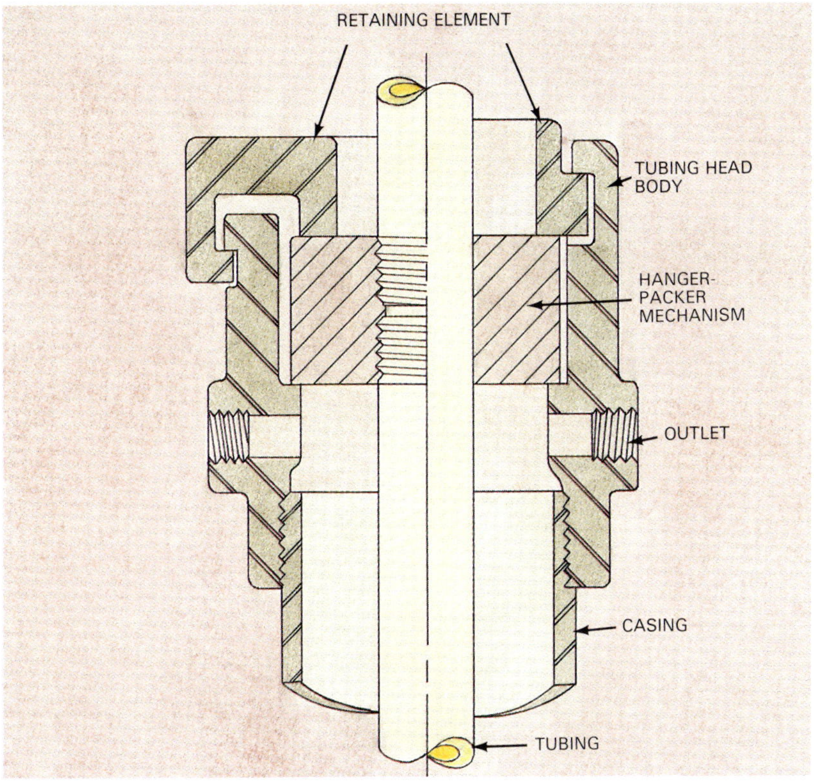

Figure 56. A tubing head can be connected directly to casing without the use of a casinghead.

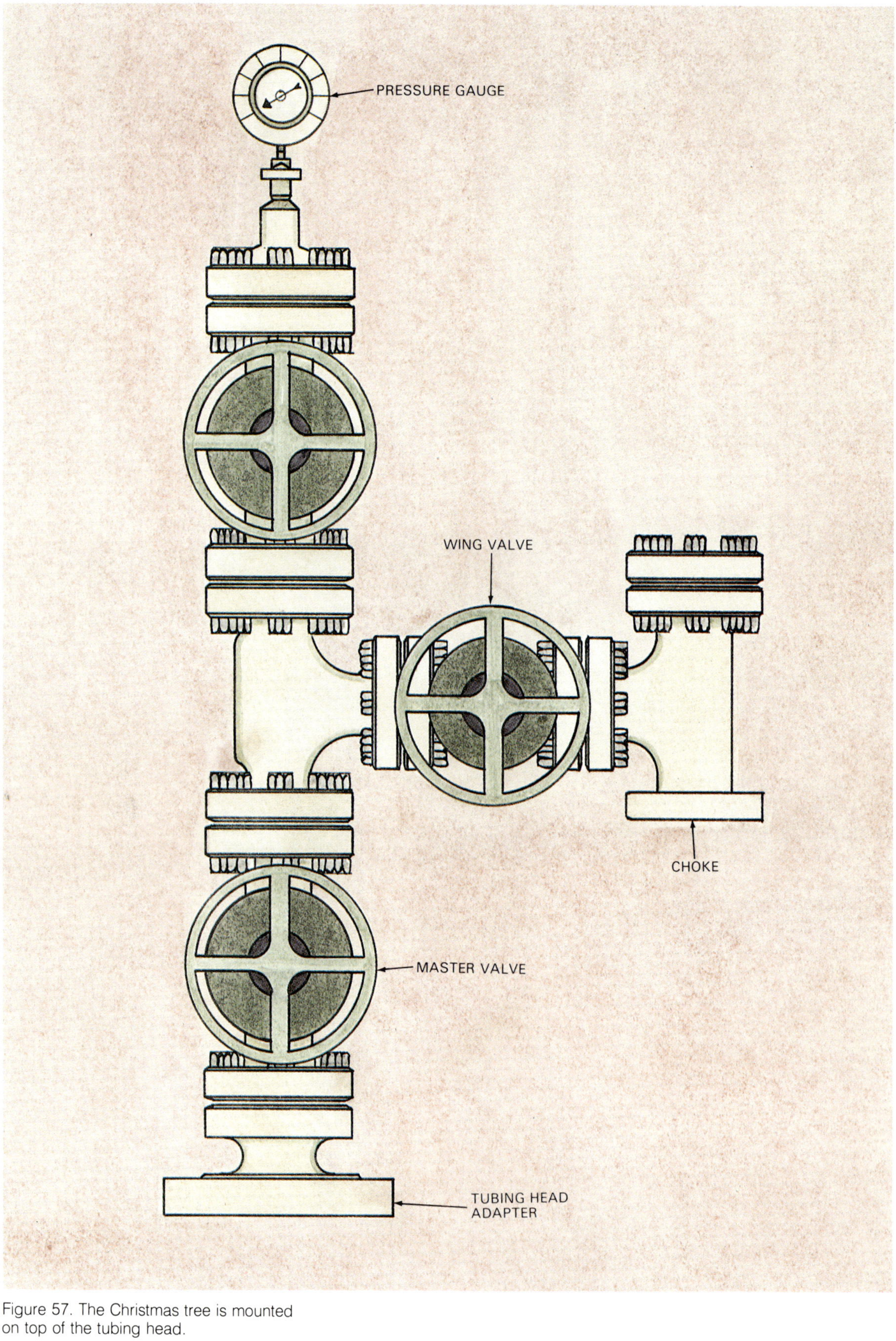

Figure 57. The Christmas tree is mounted on top of the tubing head.

Christmas Tree

High-pressure wells are usually equipped with a group of valves and fittings called a Christmas tree (fig. 57), the name derived from the treelike appearance of valves and fittings branching out from it. A Christmas tree controls the flow of oil and gas from the well and is installed on top of a casinghead or tubing head.

Low-pressure wells are also sometimes equipped with Christmas trees. These trees are less complex than high-pressure ones but serve the same purpose—the control of fluids flowing from the well.

Pressure gauges are usually installed on both high- and low-pressure trees. Such gauges measure pressure both in the casing and in the tubing. By knowing the pressures under various operating conditions, better control over the well is possible. Valves on the tree can be opened or closed to control fluid flow (fig. 58). For example, a master valve can be closed to shut off flow entirely or opened to establish flow to the rest of the tree.

Another important device in the Christmas tree is a choke (fig. 59). Basically, a choke is a restriction in the line through which well fluids exit the tree. It is used to control the production of fluids from the well. A choke may be fixed; that is, the size of its opening is permanent. If adjustments become necessary, a new choke of a different size will have to be installed. Or a choke may be adjustable; the size of its opening can be changed by turning a control handle.

Figure 58. Numerous valves are installed on the Christmas tree and can be opened or closed to control the flow of fluids from the well.

Figure 59. A choke, housed inside this fitting, controls the fluids flowing from the Christmas tree.

Artificial Lift

When pressure in an oil reservoir falls to a point where a well cannot produce by natural drive energy, some method of artificial lift must be used. There are many different types of artificial lift, but for our purposes, let's divide them into two broad categories: (1) those that employ pumps and (2) those that use gas injected into the well from the surface.

Beam Pumping

The most common method of artificial lift is beam pumping. Beam pumping consists of surface and subsurface equipment. A *beam pumping unit* is on the surface (fig. 60). It is an assembly that imparts an up-and-down motion to a string of high-strength steel rods called *sucker rods*. The sucker rods are attached to the front of the pumping unit. These rods go down inside the tubing and are attached to the top of a *sucker rod pump*. The pump is installed inside of or is a part of the tubing string near the bottom of the well (fig. 61). As the walking beam on the beam pumping unit moves up and down, the rod string also moves up and down to operate

Figure 60. A beam pumping unit is used on the surface.

the sucker rod pump. Overall, then, the beam pumping system consists of a beam pumping unit, a sucker rod string, and a sucker rod pump.

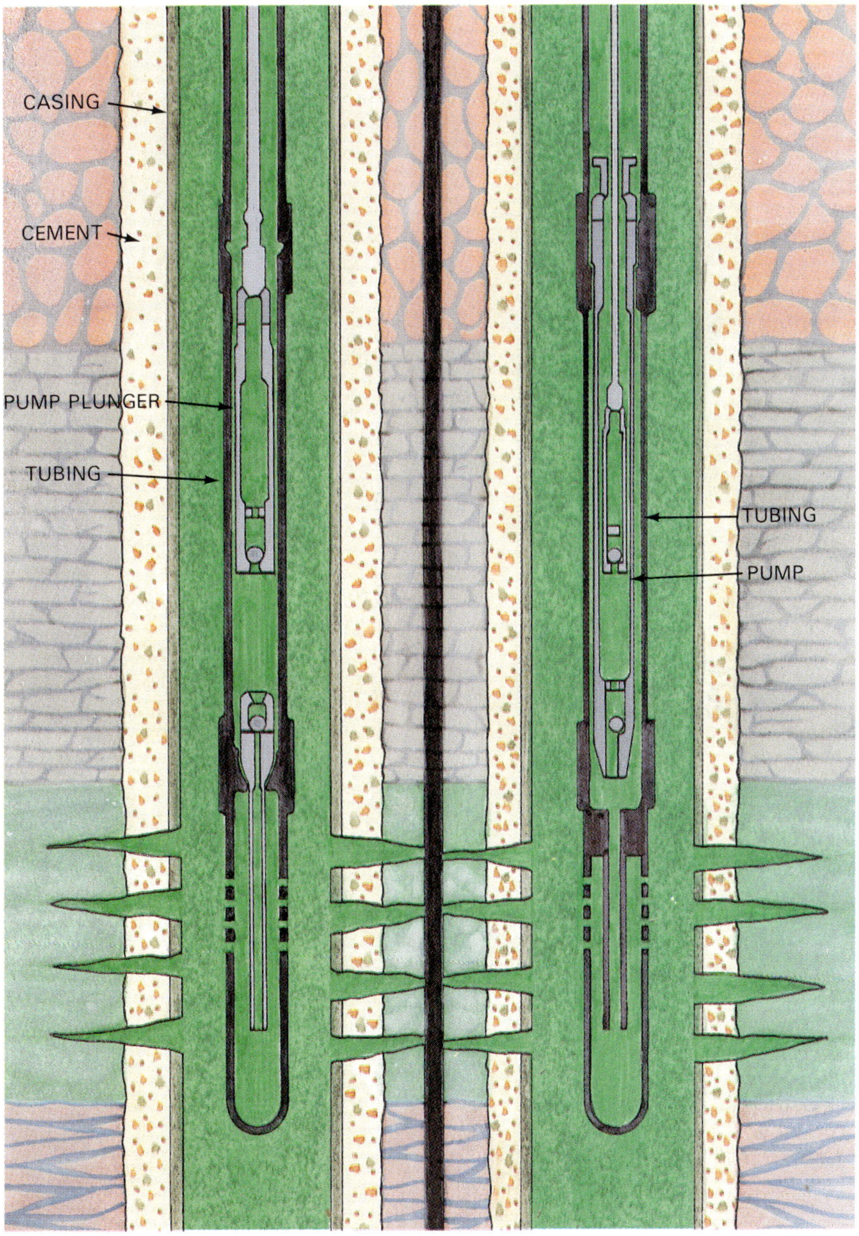

Figure 61. A sucker rod pump can be attached to the tubing string (*left*) or installed inside the tubing string (*right*).

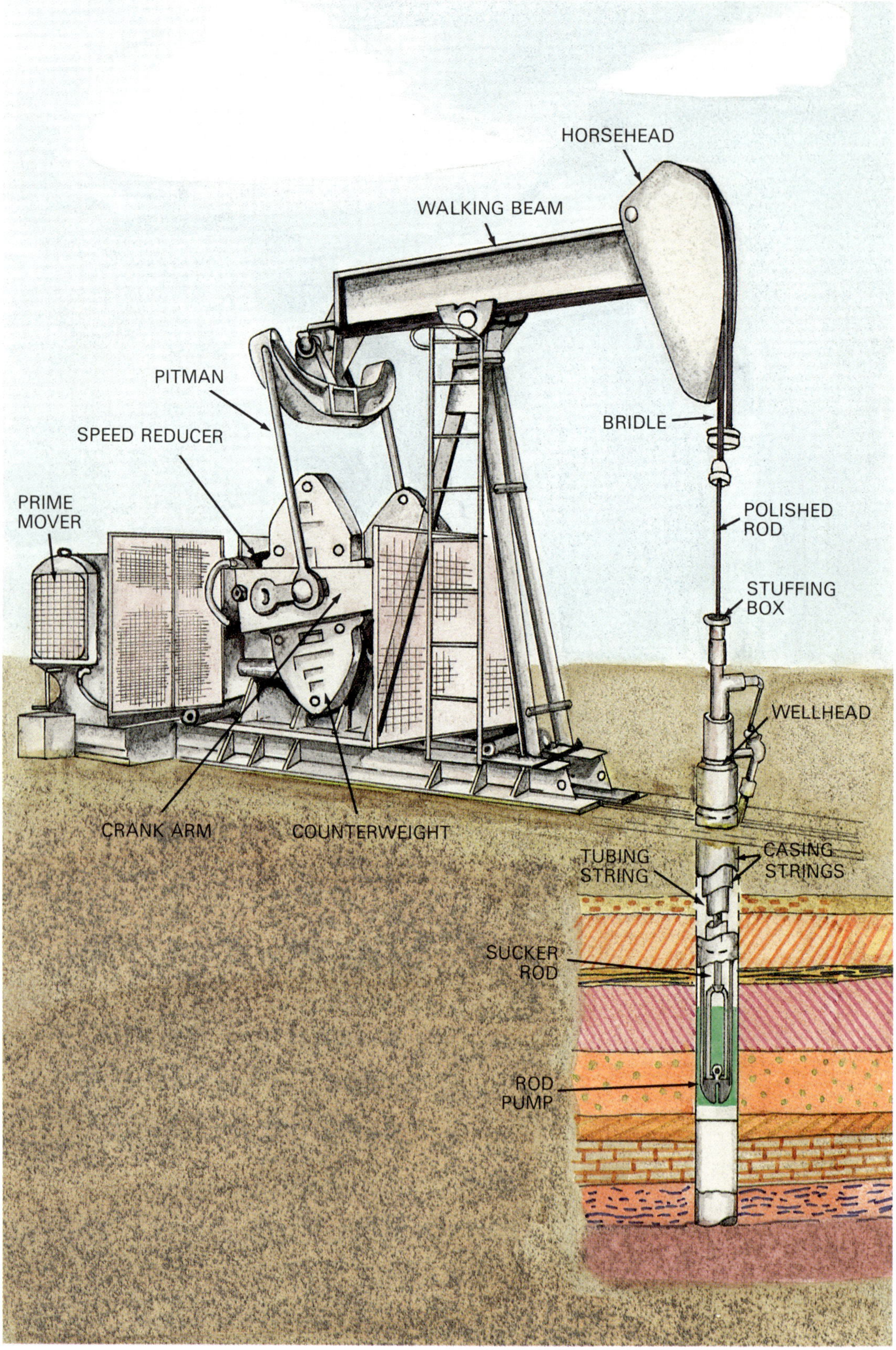

Figure 62. A beam pumping unit with wellhead, sucker rods, and sucker rod pump is one method of artificial lift.

Beam pumping units are a familiar sight around the oil patch (fig. 62). A pumping unit may be powered by an electric motor or a gas engine, usually called a *prime mover.* The prime mover (engine or motor) drives a set of gears, called a *speed reducer,* by means of pulleys and belts. The speed reducer is needed because the prime mover runs fast and the pump must run slow. For example, an engine may run at a speed of 600 revolutions per minute, but the pumping unit may need to operate at, say, 20 strokes per minute. The speed reducer makes it possible to obtain the correct pump speed.

A shaft extends from each side of the speed reducer. Connected to each end of the shaft is a steel member called the *crank arm.* A set of weights on each crank arm counterbalances the weight of the sucker rod string. (Other ways are used to counterbalance, but the use of weights is the most common.) Two metal arms, called *pitmans,* extend from the crank arms to the *walking beam.* Since the prime mover, through the speed reducer, turns the crank arms, and since the pitmans are attached to the crank arms, when the prime mover is started, the pitmans move. When the pitmans move, they impart up-and-down motion to the walking beam. The sucker rod string is connected to the front of the beam, usually with a crosspiece and two cables called the *bridle.* The bridle, in turn, is attached to the *horsehead* on front of the beam.

The first sucker rod joint that hangs from the bridle is the *polished rod.* The polished rod moves up and down inside a *stuffing box* and is usually very bright and shiny. The stuffing box has flexible material, or packing, inside of it to form a pressure-tight seal around the polished rod and the inside of the tubing. The seal keeps liquids or gas from leaking out from around the polished rod.

Sucker rods are usually solid, high-grade steel rods that are suspended inside the tubing string (fig. 63). The rods connect the surface beam pumping unit to the sucker rod pump placed near the bottom of the well. A sucker rod is

Figure 63. Several joints of sucker rods hang in the derrick of this well servicing unit prior to being lowered back into the well.

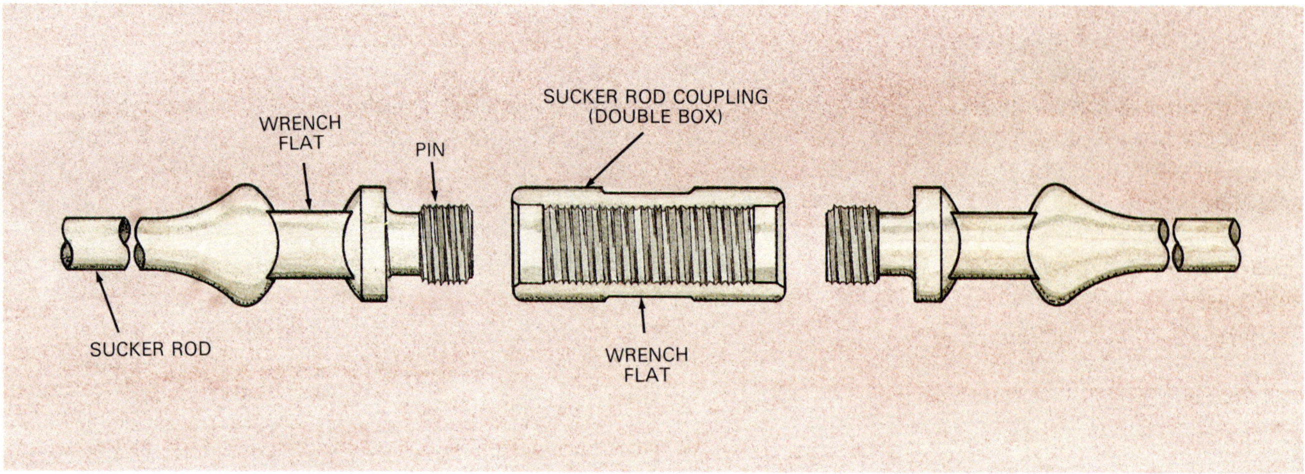

Figure 64. A sucker rod coupling connects the sucker rods.

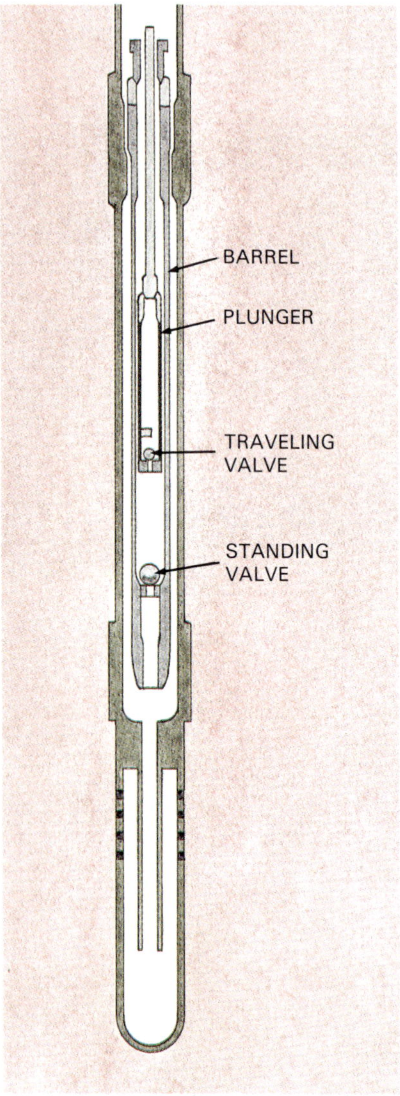

Figure 65. A sucker rod pump lifts the well production to the surface.

usually 25 feet long, so a number of rods will be connected to run from the surface to the sucker rod pump. A special coupling is used to connect each rod (fig. 64). Typically, a rod has a threaded, male connection on each end called a *pin*. Since the rods meet pin to pin, a double-box coupling is required. At every sucker rod joint, then, a coupling is used to connect the rods.

The sucker rod pump at the bottom of the rod string is used to lift the well production to the surface (fig. 65). Its parts are the *barrel, plunger, standing valve,* and *traveling valve.* A sucker rod pump works like a piston moving inside a cylinder. The plunger corresponds to the piston, and the barrel is the cylinder. The plunger is hollow and is attached to the sucker rod string. As the rod string is moved up and down by the pumping unit, the plunger moves up and down inside the barrel.

When the plunger moves up, the ball in the standing valve is pushed off its seat by the fluid below the pump to open the valve. At the same time, the ball in the traveling valve is moved against the seat by the fluid above the seat to close the valve. Fluids in the well enter the barrel through the open standing valve. On the downward stroke of the plunger, the traveling valve opens, the standing valve closes, and the fluid is forced out of the barrel through the plunger. The fluid leaves the plunger and goes into the tubing. Repeated strokes bring the fluid to the surface. At the surface, the fluid flows from the well through a pipe called a *flow line* to special facilities for processing and storing before it is sold.

Although design variations are many, two types of sucker rod pumps are basic: *tubing pumps* and *insert pumps* (fig. 66). If a well is not too deep and a lot of fluid must be lifted, a tubing pump may be used. This pump has more inside space or volume because the barrel can be the same diameter as the tubing. On the other hand, the barrel of an insert pump fits inside the tubing, so it has to be smaller than the tubing. As a result, the volume of the insert pump barrel is smaller. However, insert pumps have an advantage: to remove one from a well, only the sucker rod string has to be pulled. In a tubing pump, both the rod string and the tubing string must be pulled. Thus, insert pumps can be removed and replaced considerably faster than tubing pumps.

Figure 66. The tubing pump (*left*) and the insert pump (*right*) are two basic sucker rod pump designs.

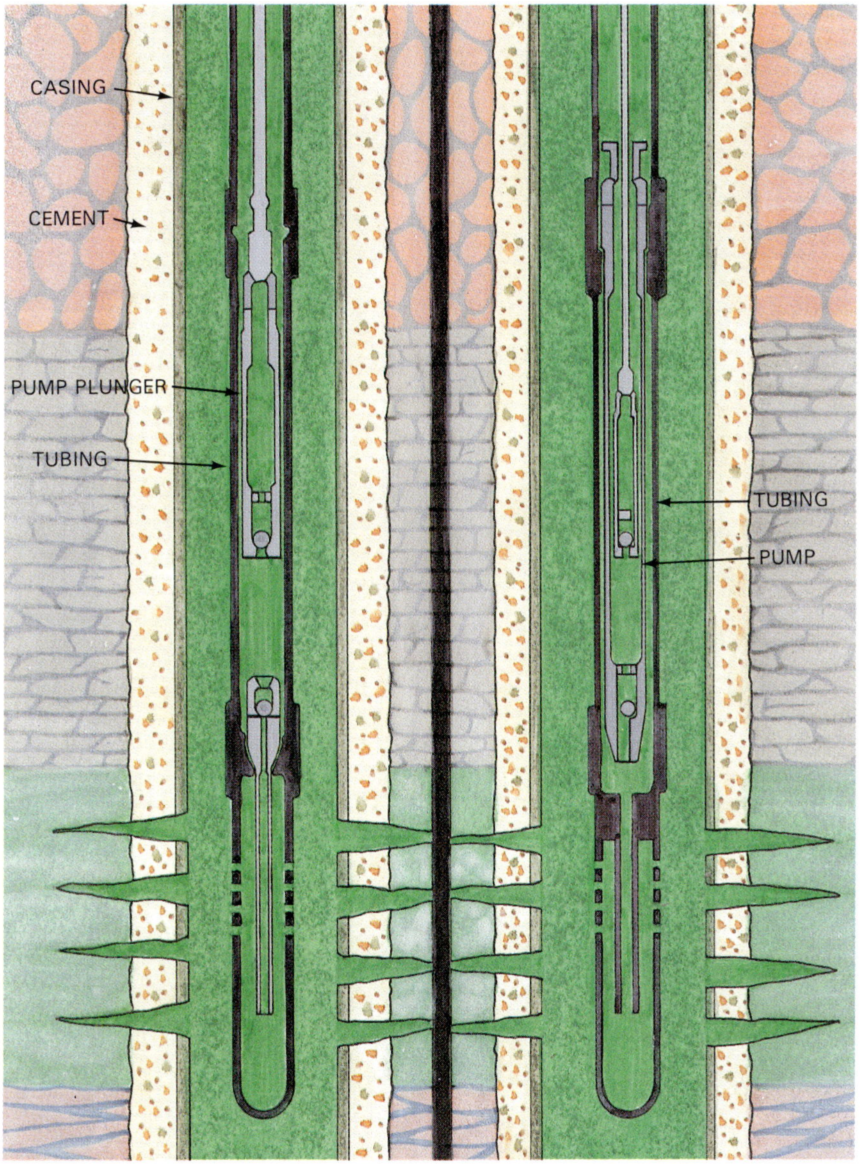

Subsurface Hydraulic Pumping

Another method of artificial lift that utilizes a pump is *subsurface hydraulic pumping.* This is a method in which fluids are lifted to the surface by a downhole pump that does not employ sucker rods. To understand how this pumping method works, it is necessary to understand the meaning of the word *hydraulics.*

Most of us are familiar with the old-fashioned waterwheel. A paddle wheel is placed in the current of a river, and the wheel turns. The turning wheel can be used to drive a machine such as a gristmill that grinds grain. In this way, energy in flowing water is put to work. This type of energy is called *hydraulic energy.* Whenever a liquid such as water or oil is put to work—turning a wheel or, in a more modern application, powering a piston in a cylinder—hydraulics is being used. A hydraulic pump, then, is a pump that is made to work by the pressure of a liquid on its pistons. The liquid is forced against the pistons and moves them. The moving pistons can, in turn, do work such as lifting well fluids to the surface.

A subsurface hydraulic pump in an oilwell is powered by the oil that the pump lifts from the reservoir. Hydraulic pumping requires a system, some of it on the surface and some of it downhole. One type of system consists of a conventional pump powered by an engine or electric motor on the surface. This pump moves oil down a tubing string in the well. The subsurface hydraulic pump is made up on the bottom of this tubing string. The oil that makes the pump work is called *power oil.* Power oil gets back to the surface through a second parallel string of tubing connected to the downhole pump. Oil and other reservoir fluids produced by the pump also move to the surface through this second tubing string.

A *settling tank* on the surface stores both the produced oil and used power oil. Any water and sediment that may be in the produced oil is allowed to settle out in the tank since the subsurface pump requires clean oil to work properly. An overall view of this type of system shows its surface and subsurface components (fig. 67).

A typical hydraulic pump consists of two pistons, one above the other and connected by a rod, that move up and down inside the pump (fig. 68). The pump is divided into an engine section and a pump section and has various passages, or ports, through which power oil and produced fluids flow. The top piston is the engine piston, and the bottom one is the pump piston. Installed in the pump section are four ball-and-seat valves that open and close as the pump moves up and down. A sliding valve in the engine section of the pump moves up and down according to the position of the engine piston.

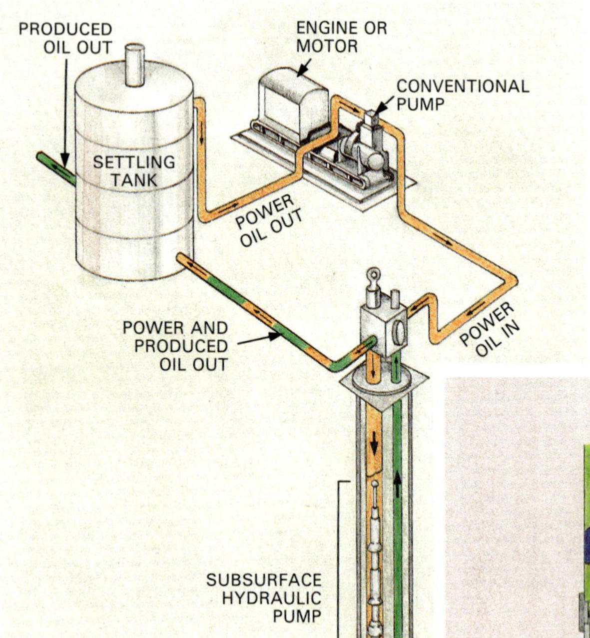

Figure 67. A hydraulic pumping system has surface and subsurface components.

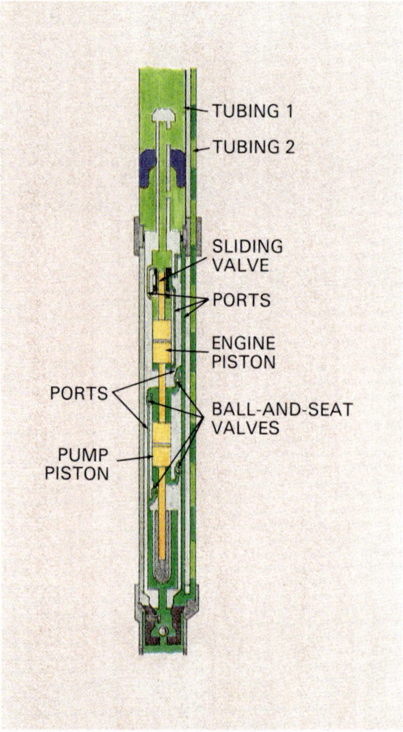

Figure 68. A subsurface, or downhole, hydraulic pump consists of pistons, ports, and valves.

Basically, the pump works like this: The surface pump sends power oil down the tubing string to the hydraulic pump. On the downstroke of the pump, the sliding valve moves down. With the sliding valve down, power oil is directed through a port to the top of the engine piston. The oil moves the engine piston and connected pump piston down. As the engine piston moves down, power oil that is under the engine piston from the previous upstroke is ported into a second tubing string and to the surface. As the pump piston moves down, reservoir fluids enter above the piston through an open ball-and-seat valve. Also, reservoir fluids under the pump piston from the previous upstroke are ported through a ball-and-seat valve into the second tubing string. These reservoir fluids join the exhausted power oil and thus also go to the surface.

On the upstroke, the sliding valve moves up. This directs power oil under the engine piston so that it moves up. Power oil above the piston is ported into the second tubing string. As the pump piston moves up, reservoir fluids enter through a ball-and-seat valve under the piston. Reservoir fluids that entered on the downstroke above the piston are ported through a ball-and-seat valve and into the second tubing string.

As the pump repeatedly strokes up and down, used power oil and reservoir fluids are moved into the settling tank on the surface. The motor or engine driven surface pump draws clean power oil off the top of the tank and pumps it back down to the hydraulic pump. Of course, some of the oil is removed from the tank and sold. The sequence of drawings in figure 69 shows in detail how one type of subsurface hydraulic pump works.

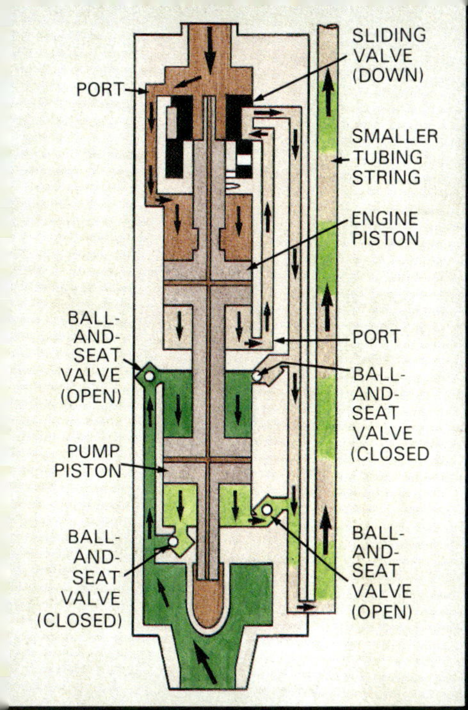

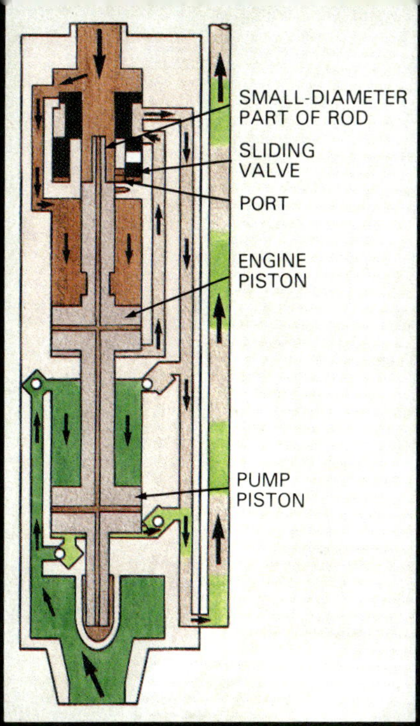

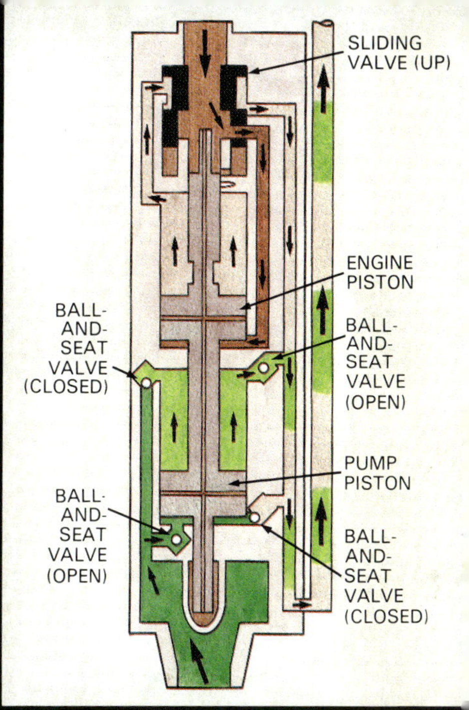

1. In the downstroke, power oil (brown) flows through a port and moves the engine piston down. Note the down position of the sliding valve. Exhausted power oil (tan) is forced out the port underneath the engine piston and enters the smaller tubing string. The pump piston also moves down and forces reservoir fluids (light green) out of the open ball-and-seat valve below this piston. These fluids join the exhausted power oil in the smaller tubing. Fluids from the reservoir (dark green) enter the space above the pump piston through the open ball-and-seat valve above and to the left of the piston.

2. With both pistons all the way down to the bottom of the downstroke, an important thing occurs. The small-diameter part of the rod at the top end allows power fluid (brown) to go around the rod, enter a port, and get under the sliding valve.

3. At the start of the upstroke, power fluid (brown) moves the sliding valve up. This opens a port that allows power fluid to get under the engine piston and start the upstroke. Exhausted power fluid (tan) above the engine piston flows through a port, around the top part of the sliding valve, and to the smaller tubing string. Reservoir fluids (light green) above the pump piston now leave through the open ball-and-seat valve to the right and above the piston. Fluids from the reservoir (dark green) enter through the open ball-and-seat valve to the left and below the piston.

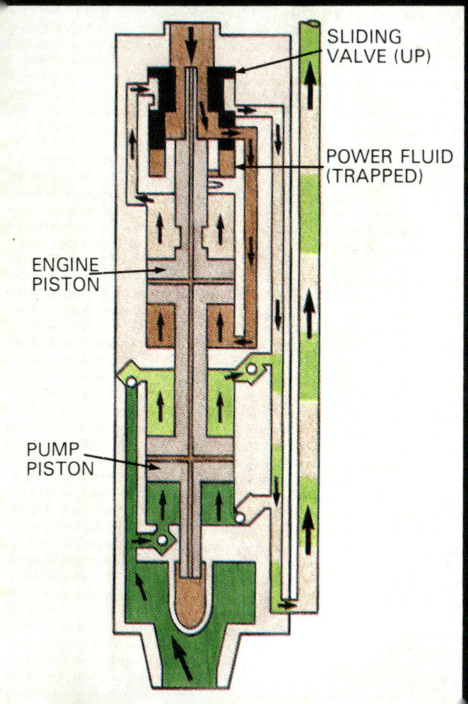

4. In the middle of the upstroke, the pressure of the power fluid (brown) trapped under the sliding valve continues to hold it up. Exhausted power oil (tan) and reservoir fluid (light green) continue to move as indicated by the arrows.

5. At the top of the upstroke, the small-diameter part of the rod just above the engine piston allows the power fluid (tan) under the sliding valve to discharge through a port to the right of the rod. When the power fluid discharges, the sliding valve moves down and the downstroke starts again.

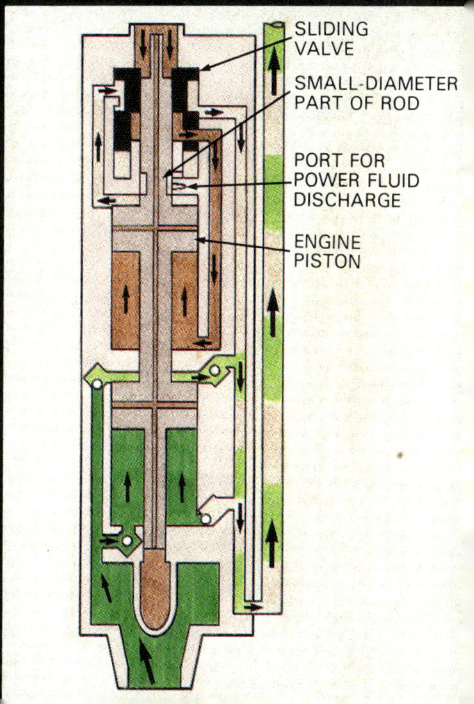

Figure 69. Schematics show how the downhole hydraulic pump works.

Electric Submersible Pumping

A submersible pump is lowered into a well until it is below the level of reservoir fluids that stand in the well—thus, its name *submersible* (fig. 70). Essentially, a submersible pump consists of a long centrifugal pump that has several *impellers* attached one above the other on a central shaft. (An impeller is simply a bladelike device that is capable of moving fluids.) This shaft is connected directly to a long electric motor. The entire assembly is placed inside the well on the bottom of the tubing string. An insulated electric cable runs from the motor to the surface, where it is connected to a source of electricity. When the motor is turned on, it spins the impellers. The revolving impellers apply pressure on the reservoir fluids that surround them. The pressure forces the fluids up to the surface through the tubing to which the whole assembly is attached.

Considerable quantities of well fluids can be lifted with a submersible pump—over 25,000 barrels per day, if necessary. In the oil patch, 1 barrel is equal to 42 gallons, so this would be over 1 million gallons per day. Of course, submersible pumps can also lift smaller quantities, but they are especially useful when it is necessary to lift large volumes of well fluids.

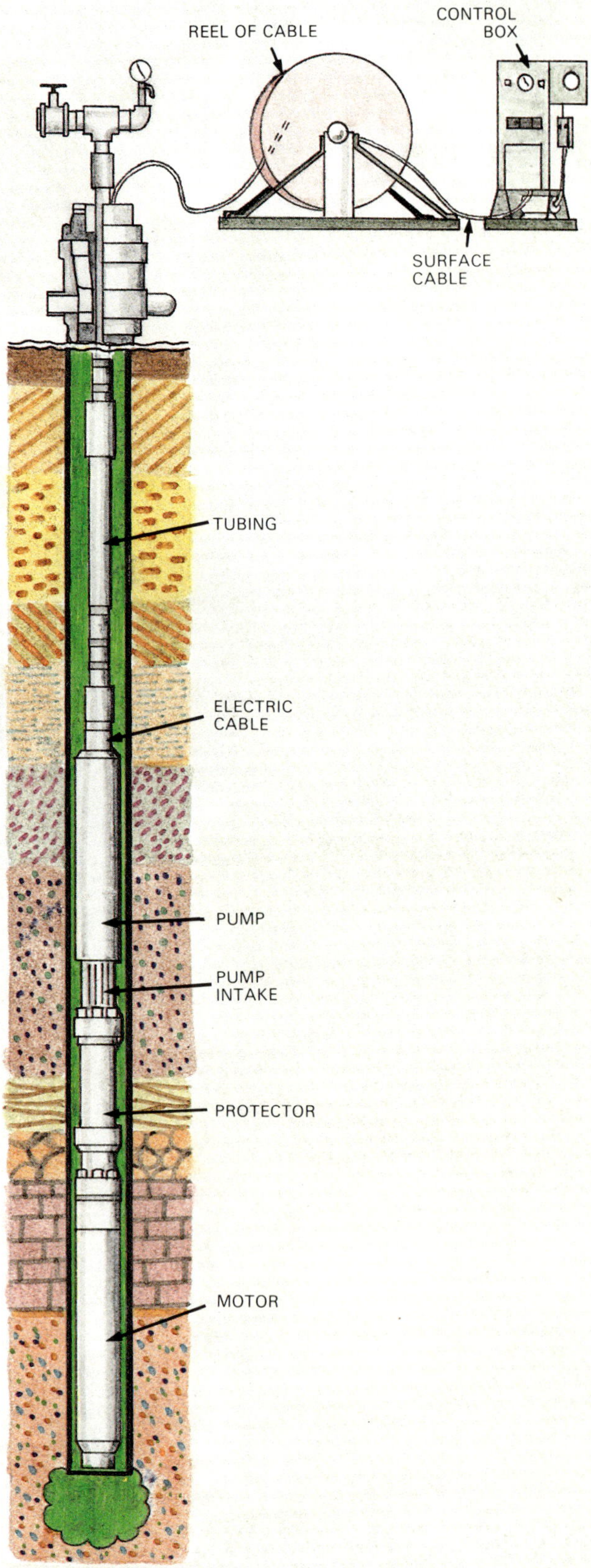

Figure 70. In an electric submersible pump, an electric motor turns impellers in the pump. The protector is an insulating section between the motor and the pump.

Gas Lift

Each of the different types of pumps is an effective way to lift oil. Indeed, probably about 80 percent of all wells on artificial lift use the beam pumping system. However, another important method of artificial lift does not use any kind of pump. Called *gas lift,* this method consists of injecting natural gas into a well in order to make it produce. Because most of the equipment used in gas lift is downhole, the method is frequently employed offshore, where space on production platforms is limited (fig. 71). However, it can be utilized anywhere, especially where a supply of natural gas is readily available.

In gas lift, natural gas is compressed and injected into the well. Usually it is injected down the casing-tubing annulus, the space between the inside of the casing and the outside of the tubing. The gas enters the tubing at various predetermined points through special valves called *gas-lift valves.* The valves open and close in a progressive manner so that the gas first enters the tubing through the highest valve and ends up entering the tubing through the lowest valve. When the gas ultimately works its way down to the lowest gas-lift valve, it has reduced the pressure in the tubing to a value low enough so that reservoir fluids can flow to the surface at an acceptable rate (fig. 72).

Figure 71. Space is limited on offshore production platforms.

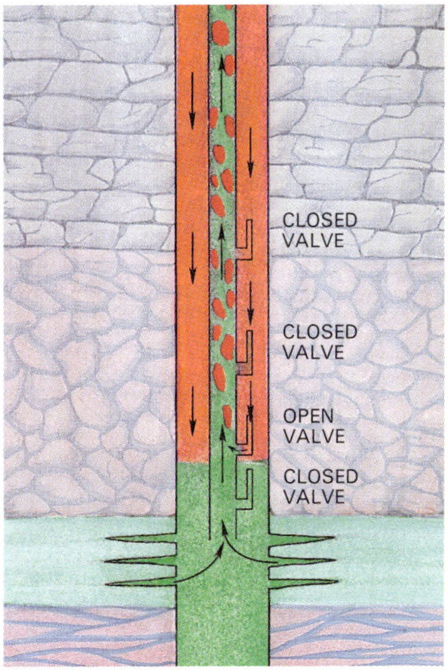

Figure 72. Gas-lift valves are installed in the tubing string at various depths. The gas enters the annulus and progresses down through the gas-lift valves.

To further understand the principle of gas lift, recall that gas is a very lightweight fluid. For example, 1 gallon of water weighs a little over 8 pounds, while 1 gallon of natural gas weighs only fractions of an ounce. When the natural gas enters the liquids in the tubing, the gas makes this liquid column much lighter in weight. Since the liquid column is lighter, it exerts much less pressure on the bottom of the well. With the pressure lower at the bottom of the well, the pressure remaining in the reservoir becomes sufficient to push reservoir fluids to the surface through the tubing.

Additional Recovery Techniques

A time comes in the life of a well when it no longer will produce oil, even when artificial lift methods are applied. This does not mean, however, that no oil remains in the reservoir. On the contrary, statistics show that from 25 to 95 percent of the original oil in place may still be there. Even the most efficient drive—water drive—can leave behind up to 75 percent of the oil. In fact, the United States Department of Energy estimates that 25 to 45 *billion* barrels of recoverable oil still remain in reservoirs that have been initially produced. This amount of oil more than doubles the current known reserves in the United States. Therefore, the petroleum industry has developed a number of additional techniques in an attempt to recover at least part of the oil still residing in many reservoirs. These techniques are known by several different names: enhanced oil recovery (EOR), secondary recovery, and tertiary recovery, to name three. Since definitions of these terms vary and overlap, no attempt is made to put the recovery methods into categories. Rather, we've put all of them under the heading of *additional recovery techniques.*

Figure 74. In waterflood projects a beam pumping unit is often used to lift the oil and water to the surface.

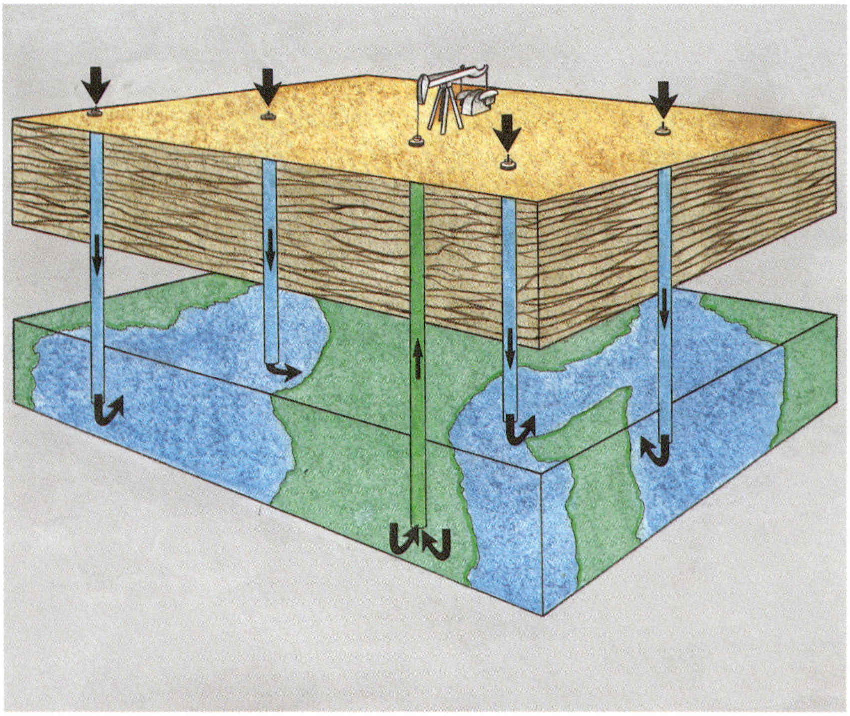

Figure 73. In waterflooding, water is injected into wells around the producing well. This is a five-spot pattern—four injection wells and the producer—but many other patterns can be used.

Waterflooding

One additional recovery technique is waterflooding. Water is injected into wells that have been drilled in one part of the reservoir, and oil and water are produced from other wells that penetrate the same reservoir (fig. 73). For example, each producing well may be surrounded by four water injection wells. (This is only one of many patterns that may be used.) In any case, water is pumped down the injection wells and into the reservoir. The water sweeps through part of the reservoir and moves some of the residual amounts of oil toward the production wells. The oil and water are then lifted to the surface (fig. 74). Waterflooding is

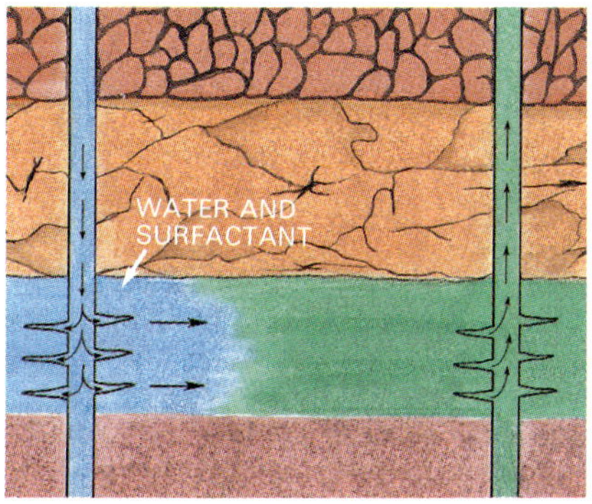

Figure 75. In one type of chemical miscible process of enhanced recovery, water containing a surfactant is injected into the reservoir.

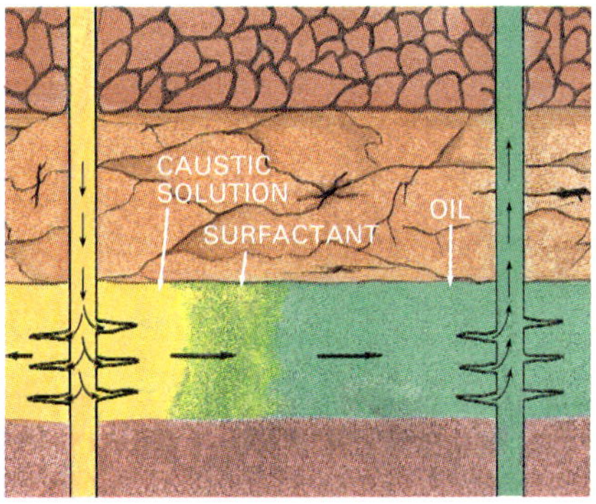

Figure 76. In another type of chemical miscible process, a caustic or alkaline solution is injected and forms a surfactant in the reservoir.

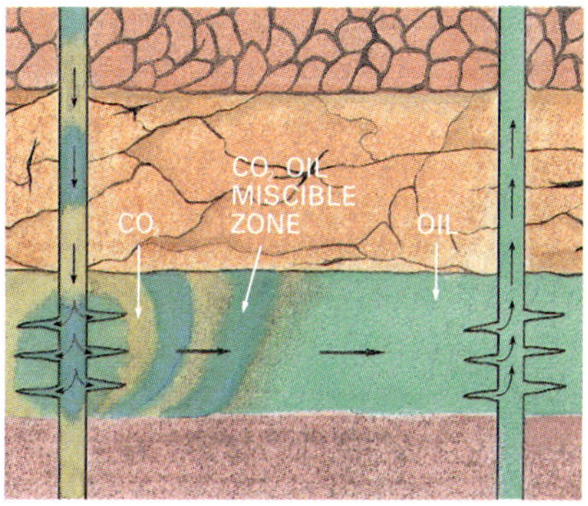

Figure 77. A kind of gas miscible process involves injecting CO_2 into the reservoir.

very similar to natural water drive and, as a result, tends to be most suitable for homogeneous reservoirs that were originally produced by dissolved-gas drive.

Miscible Processes

While waterflooding recovers additional amounts of oil from certain reservoirs, it cannot recover it all. Part of the problem is that water and oil do not mix. As a result, water will flow past some of the oil and leave it behind. Thus, two processes have been developed in an attempt to recover more oil: *gas miscible* and *chemical miscible* techniques. (Miscible simply means "able to mix.") A chemical or gas that causes water and oil to mix can be injected into a reservoir, and more oil (but not all) can sometimes be recovered.

One chemical that causes water and oil to mix is a *surfactant.* A surfactant is injected into a reservoir in two ways. A batch, or slug, of water containing a surfactant is injected (fig. 75), or an alkaline, or caustic, solution is injected. The caustic reacts with acids normally present in oil and forms a surfactant in the reservoir (fig. 76). Whether the surfactant is injected or formed in the reservoir, it washes through it (or at least part of it) and moves additional amounts of oil into production wells.

Often, a special chemical called a *polymer* is mixed with water and injected into the reservoir behind the surfactant. The polymer-water mix helps prevent channeling, or fingering, of the surfactant. Channeling leaves portions of the reservoir untouched—portions that contain oil.

A gas that is sometimes used in miscible processes is carbon dioxide, (CO_2). When CO_2 is injected into a suitable reservoir, some of the CO_2 molecules and oil molecules mix together. As CO_2 injection continues, some of the residual oil is moved to production wells (fig. 77). Usually, water is injected behind the CO_2 to minimize the tendency of CO_2 to override the lower parts of the reservoir.

Thermal Processes

Oil in some reservoirs is so thick, or viscous, that it cannot flow through the reservoir and into a well. Just as tar or other solid material can be made to flow by heating it, some viscous oils can be made to flow by heating them. Recovery techniques that use heat are termed *thermal processes.*

One method involves generating steam on the surface and injecting this steam through wells into the reservoir. When the steam enters the reservoir, it heats up the oil and reduces its viscosity. As the steam flows through the reservoir, it cools a little and hot water forms. The hot water moves the thinned oil to production wells where oil and water are produced. This process is frequently referred to as *steam flooding* (fig. 78).

Another method is *cyclic steam injection,* or *huff and puff* (fig. 79). In this method, only one well is involved, but several wells in the oilfield may have the method applied to them. Hot steam is injected down the well and into the reservoir to heat the oil. Then steam injection is stopped, the well is closed in, and the reservoir is allowed to soak for several days. In the reservoir, the steam condenses and a zone of hot water and less viscous oil forms. Finally, the well is reopened, and the hot water and thinned oil are produced. This process of steam injection, soaking, and production can be repeated until oil recovery ceases.

Still another way in which heat is applied to a reservoir is *fire flooding,* or *in situ combustion* (fig. 80). It is called in situ combustion because the technique requires that a fire be ignited in place in the reservoir. Compressed air is injected down a well and into the reservoir. The air is required to support combustion. A special heater in the well ignites the oil in the reservoir, and a fire is started. As the fire burns, it begins moving through the reservoir toward production wells. Heat thins the oil around the fire, causes gas to come out, and changes water in the reservoir to steam. Steam, hot water, and gas all act to drive oil in front of the fire to production wells.

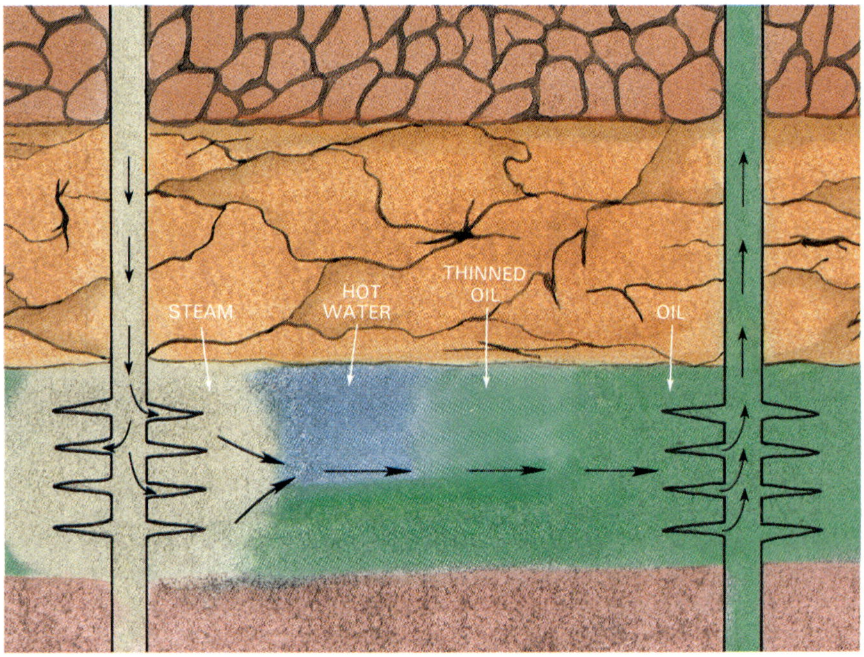

Figure 78. In steam flooding, steam and hot water thin the oil and move it to producing wells. Note that the steam and water override some of the oil, leaving it behind.

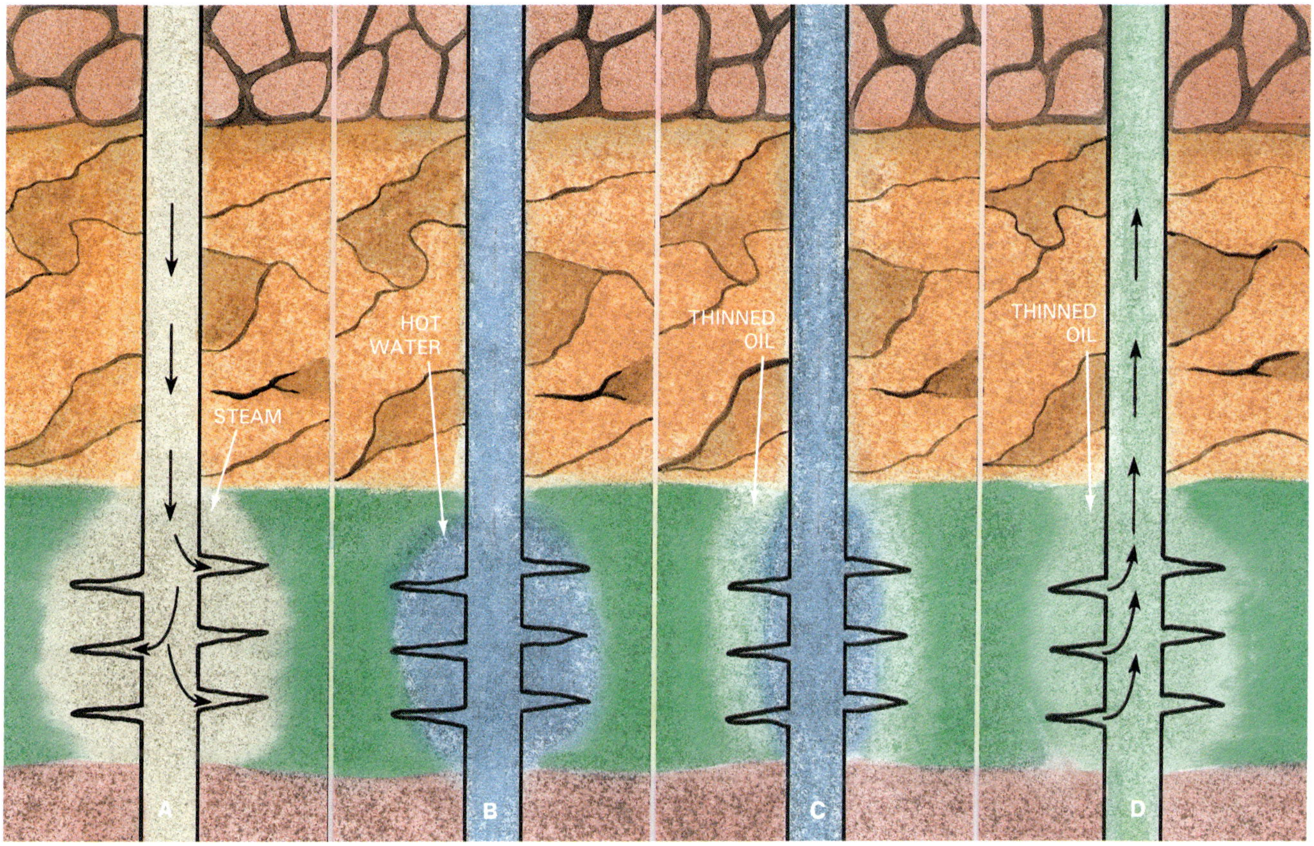

Figure 79. In the huff-and-puff method, (*left to right*) steam is injected into the reservoir; then, the well is closed in and allowed to soak; the steam cools to hot water which thins the oil; and finally, the thin oil and hot water are produced.

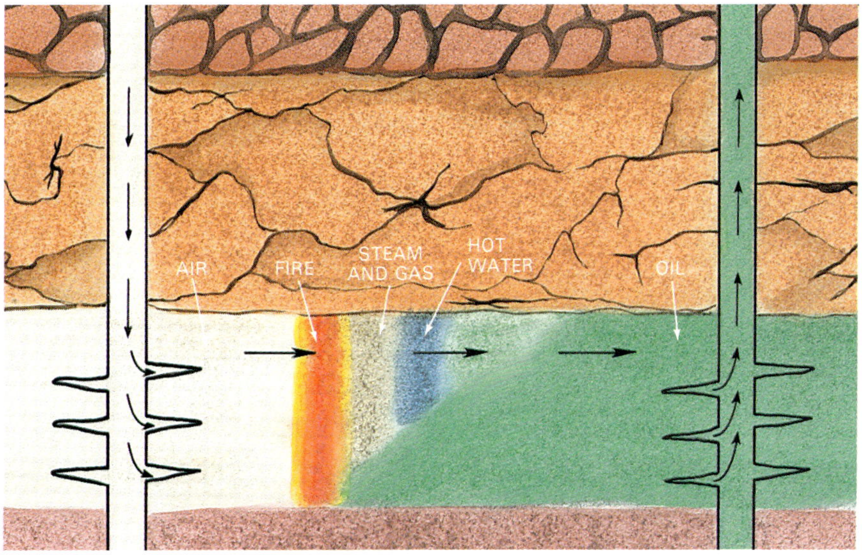

Figure 80. In the fire flood method, compressed air is injected into the reservoir, and the oil is ignited. Heat from the fire thins the oil, causes gas to come out, and turns water to steam. Steam, hot water, and gas drive the oil to the production well.

One thing to keep in mind about additional recovery techniques is that they are expensive. Special chemicals, equipment, and personnel are needed in order to initiate and maintain additional recovery projects. Further, in spite of the cost of the techniques and equipment, there are no guarantees that a project will work. Of course, the potential rewards are high if a project does work out, but risk is also high. In most cases, it takes years before a company actually starts recovering any oil from a project. Recovering oil from reservoirs beyond the initial production remains one of the great challenges facing the oil industry.

Separation and Treating

Not too much has been said about what happens to the reservoir fluids (gas, oil, and water) when they reach the surface. These fluids almost always require additional handling and processing before any hydrocarbons can be sold to a pipeline company or other shipper. Pure crude oil or natural gas seldom, if ever, come out of a reservoir. Oil can have gas and water in it; gas can have oil and water in it. Of course, the oil and gas are the fluids that are wanted Although water can be of great assistance in the reservoir, it becomes a problem when it reaches the surface.

When reservoir fluids reach the surface, the oil, gas, and water must be separated. Oil will go to a refinery where gasoline, kerosene, fuel oils, and other products will be made from it. Gas will be piped to a special plant for additional processing where products such as natural gas, propane, butane, and other hydrocarbons will be extracted. Water, which is usually salt water, has to be disposed of properly so it will not harm the environment. Often, it is injected back into a reservoir where it can help drive oil.

What comes out of a well, then, is usually a relatively complex mixture of liquid, gas, and even solids. The solids are mostly sand and scale and are called *basic sediment.* Basic sediment and water, (BS&W) have to be removed, and the oil and gas must be separated.

To further complicate matters, the water that is produced frequently occurs in two ways. Some water may be *free water,* water that will settle out of oil fairly rapidly—say, in 5 minutes or so if allowed to remain still in a tank. On the other hand, some of the water may be closely bound up in the oil. If oil and water are tightly bound together, then an *emulsion* is formed. An emulsion is a material that has droplets of one liquid spread out, or dispersed, in another liquid.

An oilfield emulsion usually has droplets of water spread out and suspended in oil (fig. 81). These water droplets will eventually settle out if given enough time; usually the process is speeded up by application of heat, chemicals, and other techniques.

In general, once the well fluids get to the surface, the problem is twofold: (1) oil and gas must be separated, and (2) water—whether it is free, emulsified with the oil, or both—has to be removed. Usually, these two operations are simply termed *separation* and *treating.*

Figure 81. This photomicrograph shows droplets of water suspended and dispersed throughout the oil in a water-in-oil emulsion.

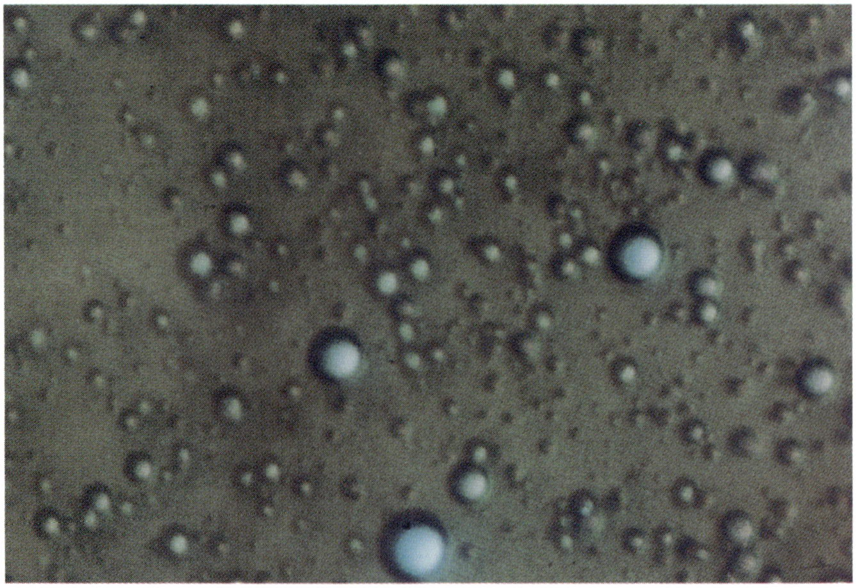

Figure 82. This vertical separator is a hollow steel vessel that receives the reservoir fluids from the well.

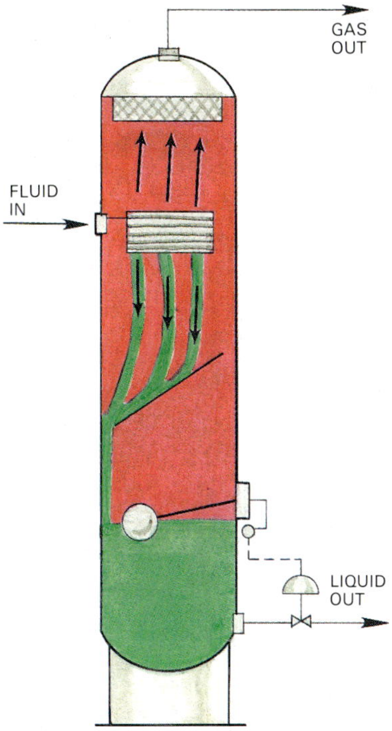

Figure 83. A two-phase separator separates well fluid into liquid and gas.

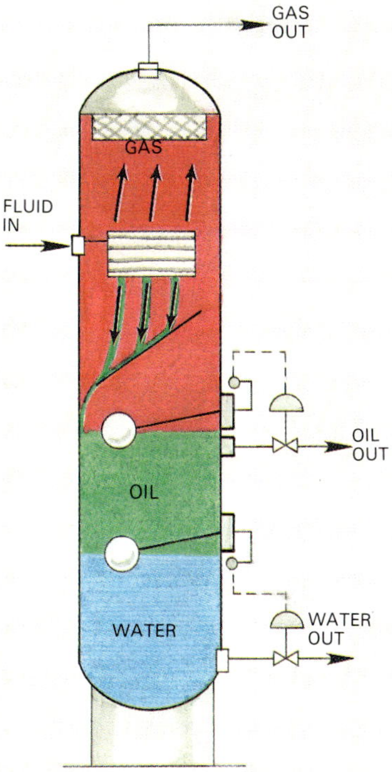

Figure 84. A three-phase separator separates well fluid into gas, oil (or emulsion), and water.

Separation

Visualize reservoir fluids flowing up the tubing, exiting the wellhead, and traveling into and through a pipe attached to the wellhead. This pipe is the *flow line*. Inside is the well stream mixture of fluids that must be separated. The fluids move along the flow line until they enter a hollow steel vessel that is called a *separator* (fig. 82). The separator allows the gas and liquids to separate from each other. The two basic types of separators are two-phase and three-phase.

A *two-phase separator* separates the well fluids into liquid and gas. The liquid—oil, emulsion, or water—goes out the bottom of the separator. The gas goes out the top (fig. 83).

A *three-phase separator* separates the fluids into oil or emulsion, gas, and water. Gas exits from the top, oil or emulsion from the middle, and water from the bottom (fig. 84). Three-phase separators may be used when there is free water in the well fluids. If there is little or no free water, then a two-phase separator may be used.

Figure 85. Basically, a horizontal separator is a tube laid on its side.

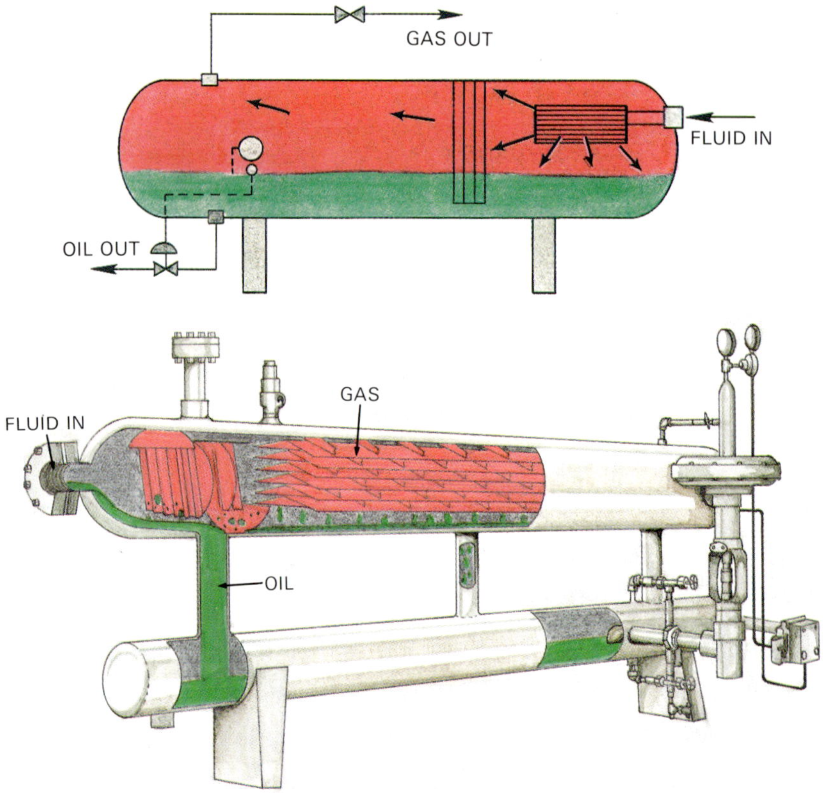

Figure 86. A horizontal separator can be single-tube (*top*) or double-tube (*bottom*).

Both two- and three-phase separators are available in different shapes and designs: horizontal, vertical, and spherical, or ball-shaped. A *horizontal separator* is one in which a cylinder, or tube, is laid on its side (fig. 85). A horizontal separator can be of either single-tube or double-tube design (fig. 86). If it is a double-tube model, the bottom tube catches the separated liquids from the top tube. In single-tube designs, all the separation occurs in one tube. *Vertical separators* are vessels that stand upright (fig. 82). Spherical separators are round, ball-shaped vessels (fig. 87). All of them do the same job: they separate well fluids.

All of the designs have advantages and disadvantages. For example, horizontal separators are generally better at separating well fluids, but they take up more space than vertical separators.

Space can be of concern when it is limited, as on an offshore production platform.

The two main functions of separators are (1) to remove liquid from gas and (2) to remove gas from liquid. To collect and remove liquid carried by the gas, a screen device called a *mist extractor* is provided. A mist extractor is designed to trap the tiny droplets of liquid as the gas passes through it. The droplets collect in the extractor and fall back to the bottom of the separator (fig. 88). To remove gas from liquid, a separator usually has flat plates called *baffles* (fig. 89). The well fluids flow over the baffles and spread out on them. Spreading out the fluid makes it easier for the gas to escape from it.

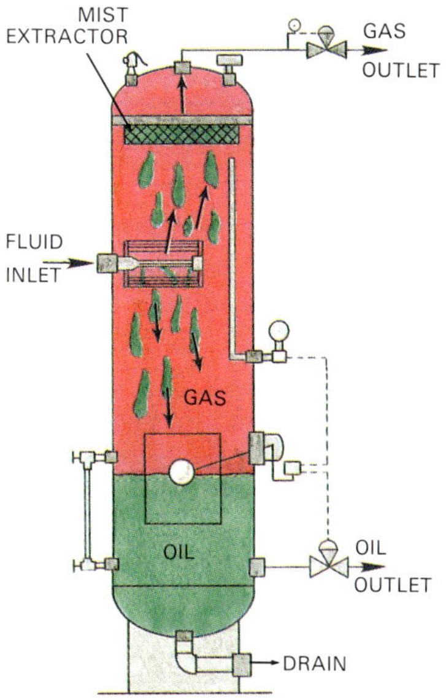

Figure 88. A mist extractor traps droplets of liquid as gas passes through it in this two-phase separator.

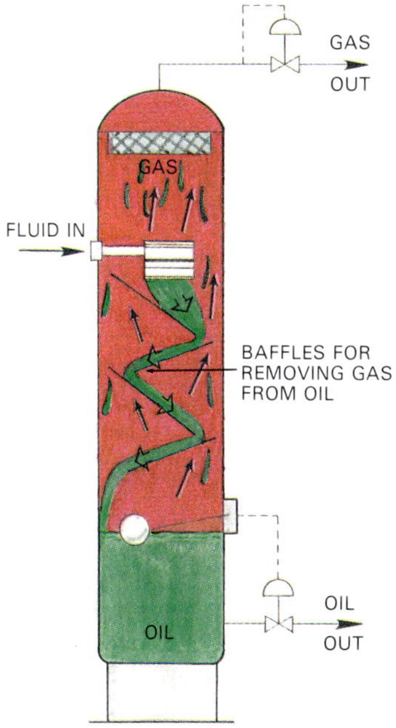

Figure 89. Baffles help gas break out of oil in this vertical, two-phase separator.

Figure 87. A spherical separator is a ball-shaped vessel, shown here on an offshore production platform.

After separation, gas will usually be measured, or metered, piped to a gas gathering system, and moved off the property to a final market. After oil is separated, it is treated and sent to one or more large steel tanks on the lease. (A lease is a piece of property on which oil and gas production takes place.) The oil is temporarily stored in tanks for sale to a pipeline company or other shipper. Oil storage tanks are called *stock tanks* (fig. 90). Water separated from the well fluids is disposed of.

Treating Oilfield Emulsions

Commonly, well fluids from an oilwell can be separated only into gas and emulsion. In other words, the separator removes gas and any free water but leaves an emulsion as the liquid. An emulsion has water droplets dispersed in the oil and usually must be *treated* in order to remove the emulsified water from the oil.

Usually, the emulsion is piped into special tanks, or vessels, for treatment. In the vessel, the emulsion will be given time to stand so that the water droplets can settle out.

Further, certain chemicals called *emulsion breakers,* or *demulsifiers,* will probably be added to the emulsion in order to make the droplets of water merge, or coalesce. When the droplets merge, they get bigger. Bigger water drops are heavier than small ones, so they settle out faster.

The emulsion may also be heated. When heated, the viscosity of an emulsion is reduced. Water

Figure 90. Stock tanks such as these are used to store oil on the lease.

droplets can fall from a thin oil faster than from a thick, viscous oil.

Electricity can also be used to break an emulsion. Passed through an electric field, water droplets take on an electric charge that makes them move rapidly. The rapidly moving droplets collide, merge, and settle out quickly.

Regardless of which of the treating methods are used, and even if all the methods are used, the emulsion must be sent to a treating tank, or vessel. Over the years, many different types have been utilized, so only a few more-modern designs will be covered here.

A heater-treater is a vessel that is a combination of a heater and other parts that aid in emulsion breaking. In *vertical heater-treaters* (fig. 91), the emulsion, usually with a breaking chemical added to it, most often flows into the vessel through a heat exchanger. Cold, incoming emulsion is preheated by warm, already treated crude oil leaving the vessel through the heat exchanger.

The emulsion then flows to the top of the treater, splashes over a pan, and falls downward through a downcomer. Gas breaks out of the emulsion as it splashes over the pan. When the emulsion reaches the lower section of the treater, any free water and sediment fall out.

Emulsion then starts rising through the water where some of the emulsified water is washed out. Also, the emulsion and water are heated by a fire tube. A flame inside the fire tube applies heat. The warm emulsion rises through the heated water and enters a settling space. There, emulsified oil and water separate. The water settles out and falls to the bottom. The clean crude oil rises upward, passes through an oil outlet to the heat exchanger, and warms incoming emulsion (fig. 92). The clean crude is piped to a stock tank.

Figure 91. A vertical heater-treater aids in emulsion breaking.

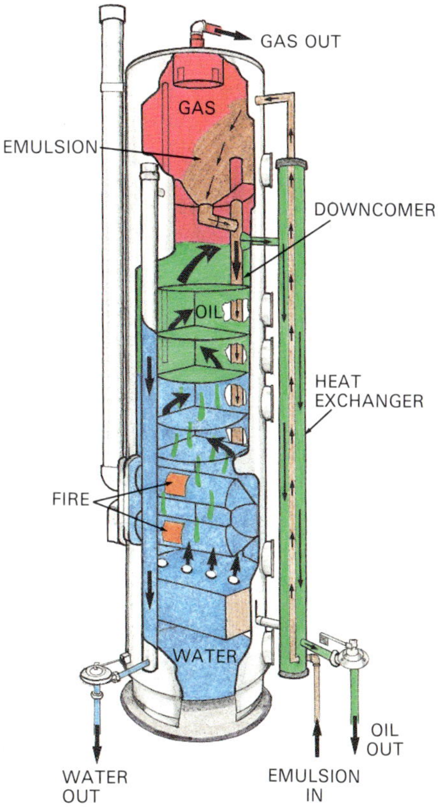

Figure 92. Components and treating action of a vertical heater-treater

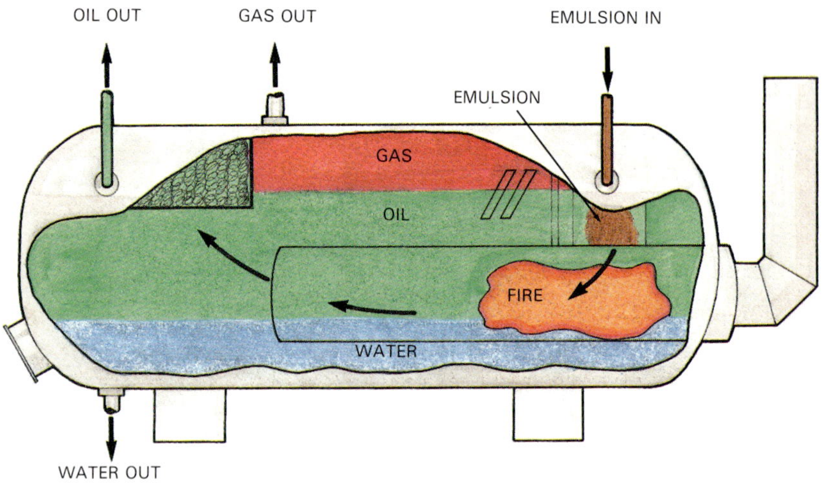

Figure 93. Components and operation of a horizontal heater-treater

Horizontal heater-treaters (fig. 93) are very similar in operation to vertical ones. Just as in vertical treaters, the incoming emulsion, with added chemical, usually passes through a heat exchanger. When emulsion enters the top of the treater, it falls downward across a fire tube. Any free water settles to the bottom.

The emulsion moves upward and crosses into a second chamber. Water breaks out of the emulsion and falls to the bottom. Clean oil and gas rise to the top.

One advantage of horizontal treaters is a large treating section, which means they can handle larger volumes of fluid than vertical designs. On the other hand, vertical treaters can handle sediment better and require less space.

An *electrostatic treater* (fig. 94) is similar in design and operation to a horizontal treater. However, an electrostatic treater features a high-voltage electric grid. The emulsion rises to the grid, and the water droplets receive a charge that causes them to collide, merge, and

Figure 94. Components and operation of an electrostatic heater-treater

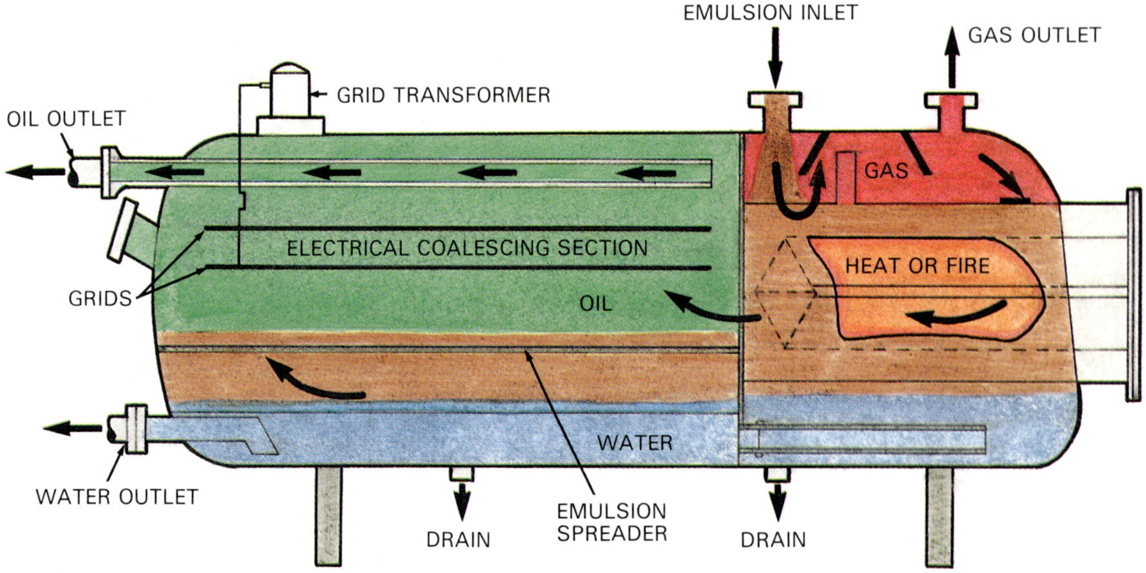

settle out. It is sometimes possible to use less heat and fewer chemicals when emulsions are treated in electrostatic treaters.

Field Handling of Natural Gas

One of the larger concerns with handling natural gas on the lease is that it usually has water vapor in it. (Water vapor is simply water in its gaseous form.) Water vapor presents a problem because it tends to readily condense back to liquid, and it can cause solids called *hydrates* to form. Liquid water frequently causes corrosion in pipes and vessels. A hydrate is a solid similar in appearance to snow and can pack tightly in a flow line and partially or fully block it.

Water vapor condenses to liquid, and hydrates form when the temperature of a gas goes down. In the reservoir, there is no problem because the temperature is warm or hot, but the temperature drops as the gas and other fluids leave the well and pass through flow lines. Since the flow lines are on or just below the surface, the gas is gradually cooled. Also, in cold weather liquid water and hydrates are more likely to form.

To combat the problem of water condensation, flow-line gas is often sent through an *indirect heater* (fig. 95). Basically, an indirect heater consists of two tubes inside a vessel. One of the tubes is a fire tube. Above the fire tube is another tube through which gas flows. This tube is often called a *tube bundle* because it is bent back on itself two or three times (fig. 96). The two tubes are surrounded by water heated by the fire tube. The heated water warms the gas flowing through the tube bundle and prevents condensation and hydrate formation.

Figure 95. An indirect heater prevents water condensation and hydrate formation in gas flow lines.

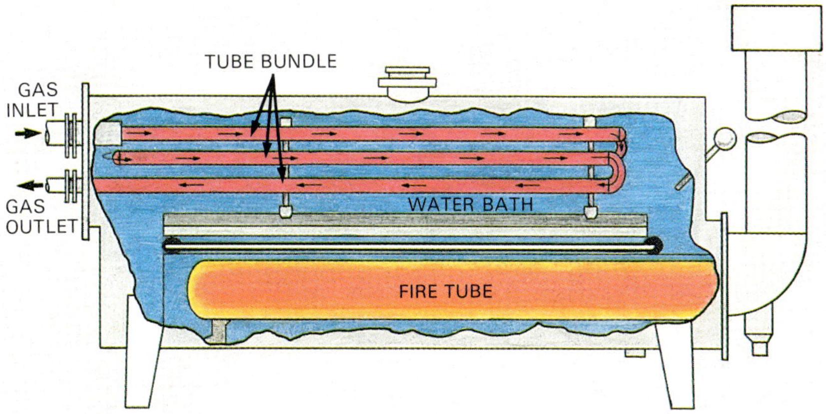

Figure 96. An indirect heater warms the water bath, which in turn warms gas flowing through the tube bundle.

Figure 97. A glycol dehydrator removes most of the water vapor from natural gas.

The problem of water vapor does not end in flow lines. Almost all water vapor must be removed before gas can be sold to a pipeline company. Hydrates and corrosion cannot be allowed to occur in pipelines for the same reasons they cannot be allowed to occur in flow lines. Removal of water vapor from gas is termed *dehydration.* Equipment used to dehydrate gas is called a *dehydrator.*

One of the more common types is the *glycol dehydrator* (fig. 97). Water vapor can be removed from natural gas by bubbling it through glycol. Glycol is a liquid that absorbs water. It is expensive—too expensive to use only once and then throw away. Thus, glycol dehydrators are designed so that the glycol can be recycled.

A typical glycol dehydration system consists of a glycol absorber, pump, surge tank, and reboiler (fig. 98). Wet gas—gas with water vapor—enters the bottom of the absorber and moves upward. Dry glycol—glycol with no

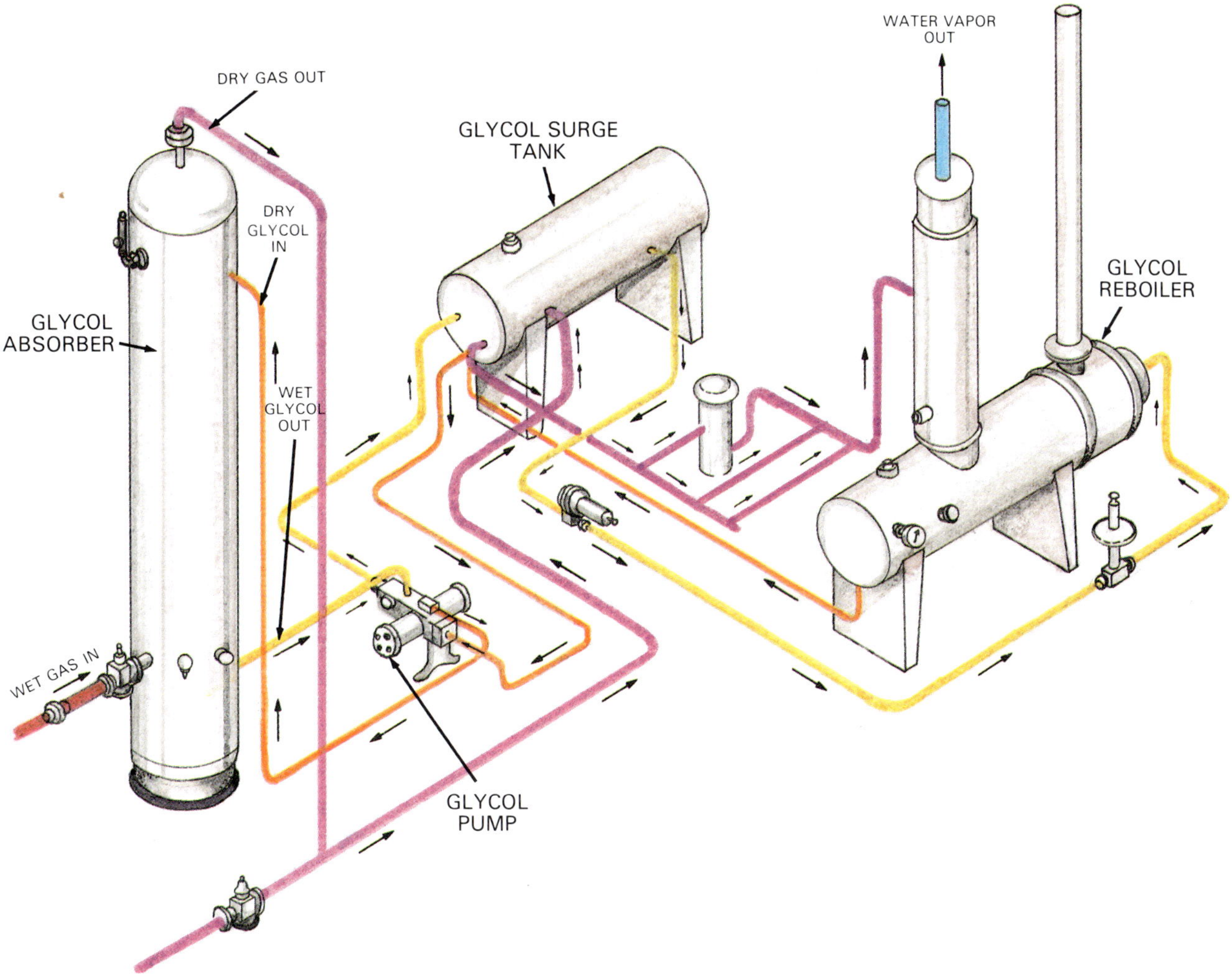

Figure 98. A glycol dehydration system consists of a glycol absorber, pump, surge tank, and reboiler.

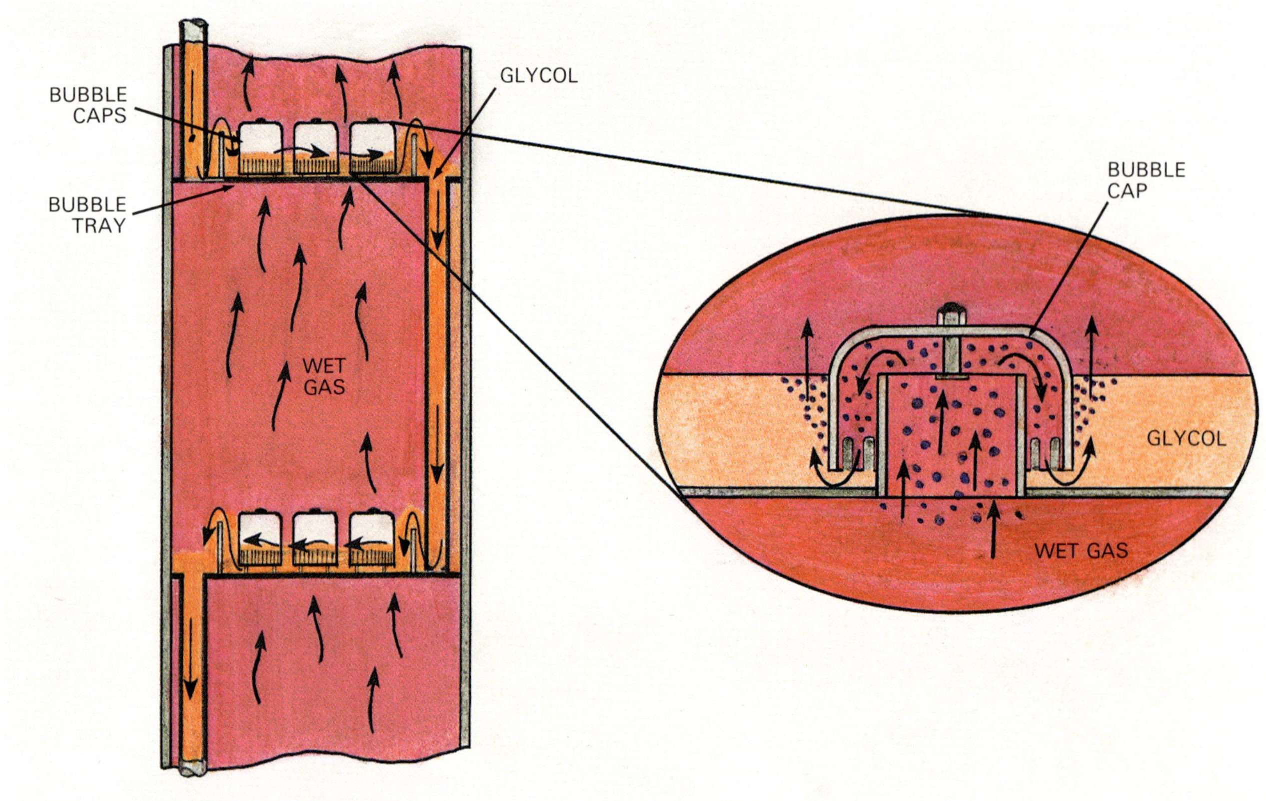

Figure 99. Bubble caps on bubble trays inside the absorber (*left*) and a close-up of a bubble cap (*right*) show how wet gas bubbles through dry glycol.

water—enters the top of the absorber and moves downward. Inside the absorber are several perforated trays with bubble caps mounted over the perforations (fig. 99). Wet gas rises through the perforations, collects under the bubble caps, and bubbles into the glycol that lies on each tray. As the wet gas bubbles up through the downward flowing dry glycol, the gas gives up its water vapor to the glycol. Dry gas exits the top of the absorber and moves to the pipeline.

Wet glycol falls to the bottom of the absorber. A glycol pump sends the wet glycol to a surge tank where it is heated a small amount. From the surge tank, the wet glycol goes to a reboiler (fig. 98). The reboiler heats the wet glycol to the boiling point of water, 212°F. Since glycol boils at about 550°F, the water boils away, leaving dry glycol. The pump then sends the dry glycol back to the absorber to complete the cycle.

Summary

Fluid coming out of a reservoir is seldom, if ever, pure gas or pure crude oil. Instead, well fluids are a combination of gas, oil, water, and sediment. The oil and gas are retained and sold, but the water and sediment must be disposed of properly.

Separators are installed to separate the fluids into gas, oil or emulsion, and water. Gas usually has water vapor in it that must be removed. An emulsion must be treated to remove emulsified water from the oil. The whole idea is for pipeline purchasers to have clean crude oil and water-free natural gas.

Well Testing

In order for oil and gas wells to be produced efficiently, information about the wells and reservoir must be obtained and studied. Much of this information is obtained from well tests.

Potential Tests

One of the well tests done most frequently is a potential test. A potential test measures the largest amount of oil and gas a well can produce over a 24-hour period under certain fixed conditions. Basically, the test involves allowing the well to produce for a given period of time and accurately measuring the production. Potential tests are performed not only when the well is first produced but throughout its entire producing life.

Well production is measured in many ways. One way is to use a *test separator* and stock tank. A test separator is simply a regular separator. However, when a well is being tested, the well fluids are diverted through the test separator in the testing system (fig. 100). In the test separator the fluids are separated into oil and gas. The gas that comes out is piped through a meter and accurately measured.

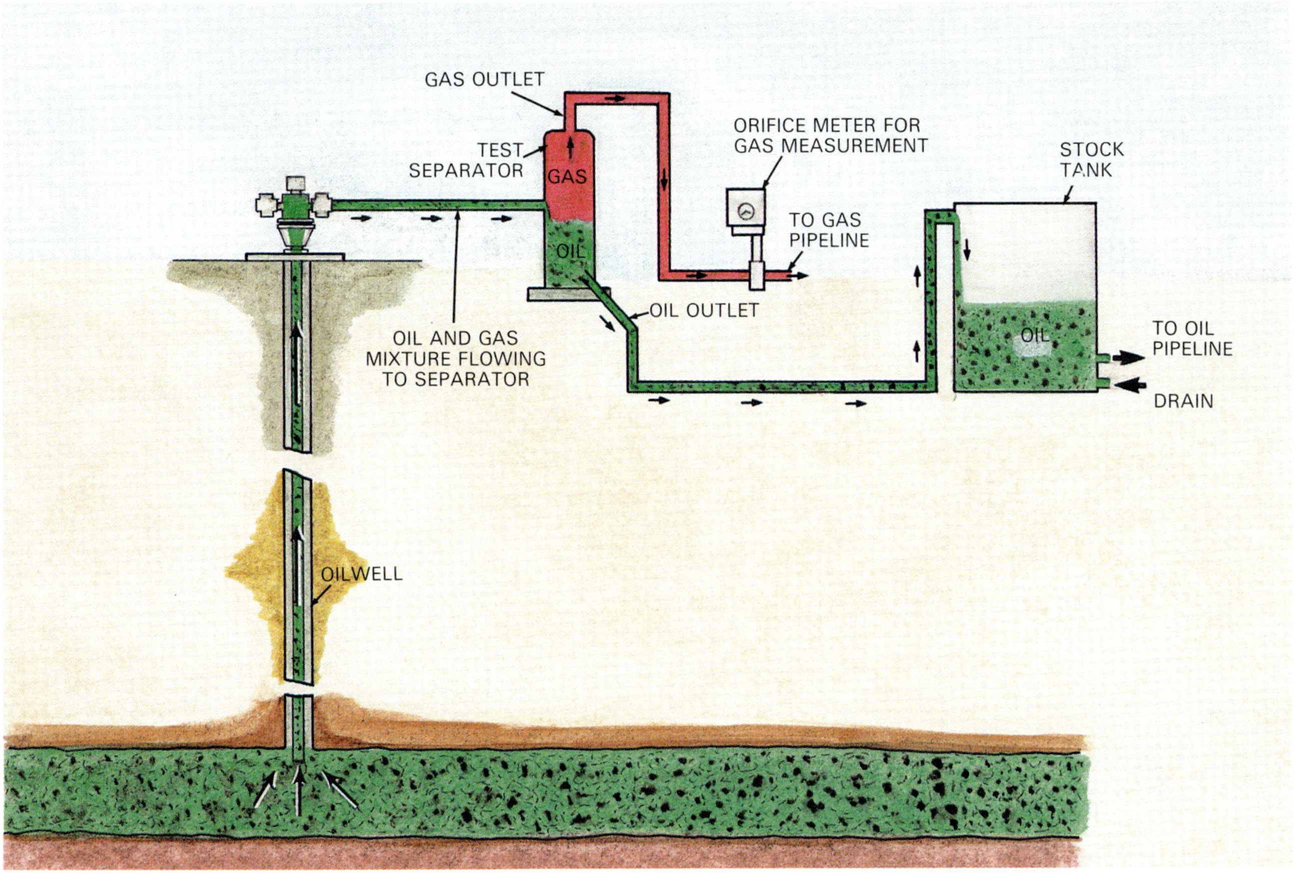

Figure 100. One type of well-testing system diverts the well fluids into a test separator where the gas is measured with an orifice meter and the oil is measured in the stock tank.

Figure 101. To measure the depth of the oil in a tank, a special flexible steel measuring tape is lowered into the stock tank.

The oil that leaves the test separator can be measured in several different ways. Sometimes, it is simply piped to a stock tank. The depth of the oil is then measured with a special tape measure (fig. 101). This measuring is called *gauging.*

Another way in which to gauge the oil is to pipe it through an automatic metering device called a *net-oil computer* (fig. 102). A special probe senses the amount of oil that flows past it. The computer electronically records the amount of oil.

Still another way in which to measure the oil is with the use of a *metering separator.* A metering separator is a test separator that has meters attached to it. It measures the oil as it leaves the separator. Metering separators are sometimes mounted on trailers so that they can be moved from lease to lease or well to well. Both net-oil computers and metering separators are generally equipped to measure salt water produced with oil also.

Figure 102. A net-oil computer is an automatic oil measurement device.

Bottomhole Pressure Tests

A bottomhole pressure test is a measure of the reservoir pressure of the well. It is obtained at a specific depth or at the midpoint of the producing zone. The purpose of the test is to measure the pressure in the zone in which the well is completed.

One way to make a bottomhole pressure test is to lower a specially designed pressure gauge into the well by means of a wireline (fig. 103). The wireline is reeled off a winch and goes through a wireline stuffing box and lubricator. The stuffing box and lubricator are designed so that the wireline and attached pressure gauge can be raised and lowered against well pressure; that is, well pressure cannot escape around the wireline. The lubricator provides a pressure-tight grease seal in the stuffing box. The gauge records the well pressure at the selected depth.

Variations of this type of test include a *flowing bottomhole pressure test,* which is a measure taken while the well continues to flow, and a *shut-in bottomhole pressure test,* which is a measurement taken after the well has been shut in for a specified period of time. Shut-in tests also give information about fluid levels in the shut-in well.

A series of bottomhole pressure tests conducted at scheduled intervals can furnish valuable information about the decline, or depletion, of the zone from which the well has been producing.

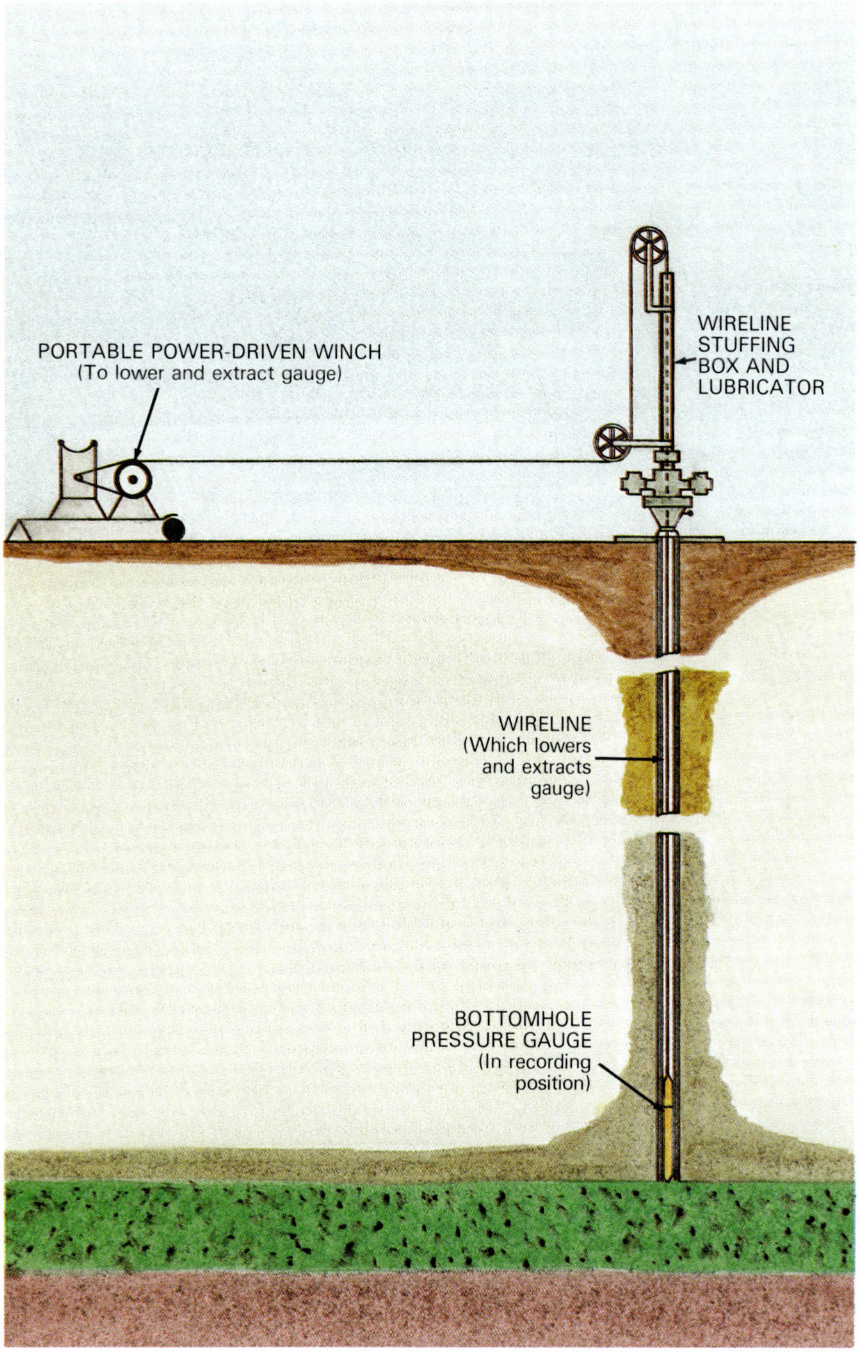

Figure 103. A bottomhole pressure gauge is lowered into the well on wireline.

Productivity Tests

A productivity test is a combination of a potential test and a bottomhole pressure test. The purpose of a productivity test is to determine the effects of different flow rates on the pressure within the producing zone of the well. The test helps establish certain physical characteristics of the reservoir. In this way, the maximum potential rate of flow can be calculated without risking possible damage to the well that might occur if the well were produced at its maximum possible flow rate.

To conduct a productivity test, the shut-in bottomhole pressure is measured first. Then, the well is opened and is produced at several stabilized rates of flow. At each rate of flow, the flowing bottomhole pressure is measured. This information, when properly interpreted, provides an estimate of the maximum flow to be expected from the well.

Special Tests

The potential test, bottomhole pressure test, and productivity test are the most common of several kinds of well tests. However, a few other special tests assist the producer. A *fluid level determination test* is most often performed on wells that are no longer flowing and must be produced with artificial lift. To help in selecting the proper type of equipment, the standing fluid level in the well is often determined. The procedure is to create a small explosion (as with a blank shotgun shell) at the wellhead. The sound is deflected by tubing couplings; by counting these deflections, the fluid level in the well can be deduced.

A *bottomhole temperature test* is normally made along with a bottomhole pressure test. The temperature test is made to determine the temperature of the well at the bottom or at some point above the bottom. To conduct the test, a specially designed recording thermometer is lowered into the well on wireline. After the thermometer is withdrawn from the well, the temperature at the desired depth is read from the instrument. The information from temperature and pressure tests can be helpful in solving problems about the nature of the oil or gas that the well produces.

Lease Storage and Measurement

Storage

Oil storage tanks on a lease are easily seen from a distance. These stock tanks are used for the storage of oil until it is sold. At the time of sale, the oil is moved by pipeline or tank truck to a refinery. Since there are usually several stock tanks, they are often referred to as a tank battery (fig. 104). The number of tanks needed depends on how much oil is being produced and how often the oil is sold to a purchaser. Generally, there are enough tanks to store anywhere from three to seven days' worth of production. Usually, at least two tanks are used so that while oil is being drained from one, the other can be filling.

Figure 104. A top view shows three tanks in this tank battery.

Figure 105. An oil sales outlet is located near the bottom of each stock tank.

Near the bottom of each tank is an important drain, or outlet. This outlet is used to take oil out of the tank for sale to a pipeline company or other shipper. The sales outlet is located about a foot or so from the bottom (fig. 105). The tank space below the oil outlet allows room for any basic sediment and water that may still remain in the oil to settle out.

The *oil sales outlet* has a special valve on it. When the tank is being filled with oil, the valve is closed and a metal seal clamped to it. When the tank is being emptied, the valve is sealed open. Sealing the valve open and closed assures both the oil producer and shipper that only oil from a particular tank will enter the shipper's pipeline or truck.

Generally, oil from a separator or treater is piped into the tank near the top. Usually, a valve on the inlet line can be closed when the tank becomes full. On top of the tank is a *thief hatch* (fig. 106), which is simply an opening with a lid. A steel tape measure is lowered and raised in the tank through the hatch to determine how much oil is in it. The hatch is also used as an opening into which a *thief* can be lowered. A thief is simply a small metal and plastic cylinder that is lowered into the tank to obtain, or "thief," a sample of oil (fig. 107). The metal used for the cylinder is usually a nonsparking one such as brass.

Oil samples are needed so its BS&W content and weight, or gravity, can be measured. Why does BS&W content and gravity of oil need to be measured? The answer brings us to the measurement, or gauging, aspect of handling oil.

Figure 106. A thief hatch (shown open) is located on top of each tank.

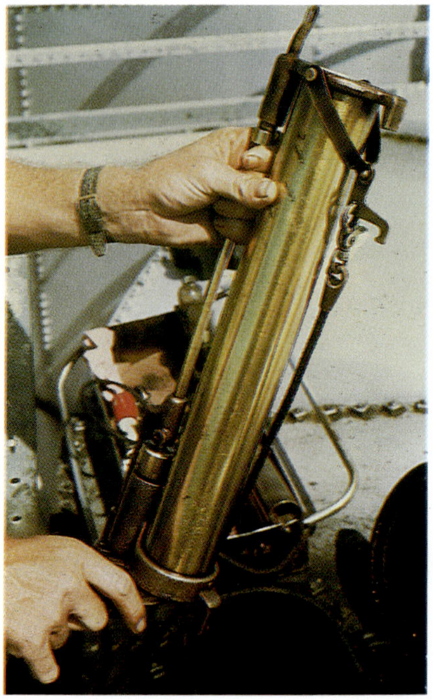

Figure 107. A thief is a special calibrated cylinder used to get a sample of oil from the tank.

Measurement

In the United States, crude oil is bought and sold by the barrel, and 1 barrel is equal to 42 gallons. Of paramount importance to the seller and buyer is knowing precisely how many barrels of oil a tank contains. Knowing how many barrels of oil is in a tank may not be as easy to determine as we might think. For instance, almost always there is a little water and sediment in the oil in the stock tank. The buyer is going to pay for the oil only, not the water and sediment. Therefore, it's necessary to determine how much BS&W is in the oil so that its volume can be deducted from the total volume of fluids in the tank.

Also affecting the volume of oil is its temperature. The warmer the oil, the more volume it occupies;

Figure 108. A LACT unit automatically transfers crude oil from the lease to the pipeline.

conversely, the cooler the oil, the less volume it occupies. A standard temperature has been established upon which volume measurements are based. In the United States this standard temperature is 60 degrees Fahrenheit. In other words, regardless of the actual temperature of the oil, its volume is computed as though it were always at 60°F.

Further, all crude oil does not sell for the same price. Usually, the less the crude oil weighs (the lighter it is), the more it sells for.

In the oil patch, the weight of oil is expressed in *API gravity.* API stands for the American Petroleum Institute, a petroleum trade and standardizing organization with headquarters in the United States. Gravity is simply the weight, or density, of the oil. API gravity is expressed in degrees. For example, an oil may have a gravity of 40°API. Incidentally, the higher the API gravity, the lighter the oil. So, a 40°API oil is lighter than a 30°API oil. The temperature of the oil also affects its gravity. Once again, a standard of 60°F is used as a base for all gravity measurements in the United States.

When oil is sold to a shipper, more than simply draining the tank is involved. First, any water and sediment are drained. Then, the oil in the tank is gauged; that is, its volume in barrels is measured. Next, the average temperature of the oil is measured, as is its gravity and BS&W content. Then, the tank is drained into the pipeline or tank truck. (It will drain down to the level of the sales outlet near the bottom of the tank.) Finally, the tank is gauged once more to determine the volume left in the tank below the sales outlet. Both the well owner's representative and the shipper's representative will witness all of these measurements.

LACT Units

Equipment is in use that automatically takes crude oil samples, records the temperature, determines the quality and quantity of the oil, shuts down and sounds an alarm if it senses that something has gone wrong, and performs other tasks that must be done to successfully transfer oil from lease stock tanks to a pipeline. The equipment that does all these operations is called a *lease automatic custody transfer unit,* or *LACT unit* for short (fig. 108). It is called this because it automatically transfers crude oil from the lease to the pipeline.

A LACT unit can be set up so that all the crude oil from the tank battery is piped through it, measured and sampled, and then put into the pipeline. A LACT unit can also send bad oil (oil that has

Figure 109. Computers are used in many modern production operations.

too much BS&W in it to be acceptable by the pipeline) back to the treating system for further cleanup. In short, a LACT unit gauges, samples, and tests all the oil on the lease as it flows into the pipeline. And it automatically transfers the ownership, or custody, of the oil from the lease owner to the pipeline.

Computers in Producing Operations

Computers—our modern electronic devices that perform repetitious calculations—are being used more and more in modern production operations (fig. 109). Computers are relatively easy to use, but skilled persons are needed to keep them functioning properly.

Computers can control and monitor oil, gas, and water production. Computerized electronic controls can operate LACT units and automatic well testing systems. Also, they can control the flow of wells, the storage of oil, and alarm devices. They collect and record production information such as flow rates from wells, liquid levels in stock tanks, pressures, temperatures, and so on. In short, computers can do routine work and keep a 24-hour watch on nearly all operations. Computers allow the people working on the lease more time to perform other equally important duties that ensure uninterrupted production.

Summary

In summary, oil that is produced from wells on a lease must be sold to a shipper. The shipper is responsible for transporting the oil to a refinery. Since the oil that comes out of wells is not pure crude oil, it must be treated. Any gas must be separated from it, and most of the water and sediment must be removed. After treatment, it has to be routed to the tank battery for temporary storage. The oil must be stored for a time so that it can be tested and accurately measured, or gauged, for sale to the shipper. Lease automatic custody transfer (LACT) units are often used to automatically gauge and test the oil as it is put into the pipeline. Finally, computers have helped make lease operations much more efficient.

Special Problems

The routine handling and treating of well fluids on a lease is a continuous operation, and problems always seem to develop. One of the biggest problems is the water that almost every well produces. This water is usually salty, and most of us know what salt water can do to steel—turn it into a rusty hulk in a very short time. Rust is a type of corrosion. Corrosion of steel components on a lease is a constant concern.

Another problem associated with water production is the formation of scale. Minerals in water can come out of it and leave a deposit inside pipe and other lease equipment. Scale can plug pipe and cause other problems. Still another problem with water is that it must be properly disposed of.

The oil and gas produced are seldom problem-free either. Sometimes a solid substance called *paraffin* comes out of the reservoir with the oil and blocks lines. Oilwell paraffin is a brown or black sticky substance that can adhere to the inside of the tubing in the well and plug it.

Oil and gas can have a sulfur compound in them called *hydrogen sulfide,* or H_2S. H_2S is a colorless gas and, in small amounts, smells like rotten eggs. Oil with H_2S in it is called *sour crude.* Gas with H_2S in it is called *sour gas.* Sour crude and gas can be harmful to equipment and, even worse, can be hazardous to people. H_2S is very poisonous in concentrated amounts. Also, in the presence of water, H_2S is very corrosive to ordinary steel.

Corrosion

Corrosion is one of the major problems for the oil industry. As oilfields get older, more and more water is produced. Increased water production means increased corrosion problems. One of the best ways in which to fight corrosion is with special chemicals called *inhibitors.* Through research, industry has developed inhibitors that give good protection against corrosion at a reasonable cost. Most corrosion inhibitors are in liquid form, and they at least hold down the corrosion rate if not get rid of it completely.

The best place to put a corrosion inhibitor is in the well fluids as they come out of the reservoir. There the inhibitor mixes with the fluids and goes to work immediately at the bottom of the well. Then, as the fluids flow up the tubing and into the rest of the production system, the inhibitor is already in the fluid stream.

Frequently, a chemical injection pump is used to inject liquid inhibitor into the well (fig. 110). In cases where it is not possible to pump

Figure 110. A chemical injection pump is often used to inject corrosion inhibitors into the well stream.

Figure 111. Scale deposits can decrease the flow capacity of pipes.

inhibitor, then a solid inhibitor, or stick, can be dropped down the well. The stick dissolves, or melts, and mixes with the well fluids as they go up the well.

Special steel alloys are also helpful in the battle against corrosion. Alloys are composed of various metals that naturally resist corrosion. Oilfied equipment made from these steel alloys has a high resistance to corrosion.

Plastics and plastic paints are other helpful corrosion fighters. Plastic coatings inside of pipes and tanks protect them from corrosion. Also, in low-pressure systems, plastic pipe and plastic or fiberglass tanks may be used. However, plastic usually does not have the strength to work in high-pressure systems.

Some types of corrosion are at their worst when air is present in the production system. (Actually, the oxygen in the air is the culprit.) Therefore, one way in which to keep corrosion down is to *eliminate air* in pipes and tanks as much as possible. Often, air can be kept out simply by repairing leaks and maintaining the level of fluids in a tank as high as possible.

Cathodic protection is another way in which corrosion can be fought. Whenever pipe is placed in the ground, corrosion will start in certain places along the outside of the pipe. This corrosion is caused by weak electric currents flowing between the pipe and soil. Basically, what happens is that electricity flows away from the pipe and into the ground. As the electricity flows, it carries very small parts of the iron in the pipe with it. Given enough time, the small parts add up to a large spot of corrosion.

To combat this type of corrosion, a bar of metal (like magnesium) can be buried near the pipe. Now electricity will flow away from the magnesium and toward the pipe. The magnesium slowly corrodes away instead of the pipe. This is called cathodic protection.

Scale

Scale is a mineral deposit that comes out of water and sticks to anything the water happens to be in. Scale inside of pipes restricts and even stops the flow of fluids (fig. 111). Scale inside a heater cuts down its efficiency because the scale has to be heated before the water can get hot.

One way to handle scale is with chemical treatment. A good scale-inhibiting chemical will either prevent scale from forming or prevent scale that does form from sticking to the pipe or equipment. Some scale is easily removed by putting acid into pipes or tanks. The acid dissolves the scale. However, some scale does not readily dissolve in acid. In such cases, the scale may have to be sandblasted or scraped away. In extreme cases, the pipe or equipment may have to be replaced.

Water Disposal

The proper handling of produced water requires special care. Water that comes from an oilwell is usually salt water, or brine. The brine may contain trace amounts of oil as well as scale. The best way to handle this water is to provide some settling time in a tank where air is excluded. Then, if wells are available, the brine can be pumped back into the producing reservoir.

Injection of water into a reservoir requires special equipment such as water injection pumps, filters, and water storage tanks. Care must be taken to avoid severe corrosion and scale problems. However, injection systems are being used more and more. Nearly all water produced with oil and gas is disposed of by injection back into the producing reservoir or into reservoirs containing only salt water.

Paraffin

Paraffin is actually oil, but it is oil in its solid form and can be troublesome. Paraffin is a waxy substance deposited by crude oil as it flows up the well and through the surface equipment. The main cause of paraffin deposits is the cooling of the oil stream as it flows from the reservoir to the surface. Under reservoir temperature conditions, the paraffin is liquid. When the well fluids flow to the surface where it is cooler, paraffin turns into a solid.

Even though research continues, a chemical that is consistently effective in preventing paraffin formation has not been found. Thus, about all that can be done is to remove it after it forms. Paraffin can be removed in several ways. One way is to pump a solvent into the flow line and tubing where it dissolves the paraffin. Solvents are expensive and often hazardous to use; therefore, their use is limited.

A more common way to get rid of paraffin is to inject hot oil down the tubing where it melts the paraffin. Steam is also used in the same way. On pumping wells, where the tubing has sucker rods in it, injecting anything into the tubing is difficult. So, sometimes, hot oil or steam is injected into the annulus (the space between the tubing and the casing). This softens the paraffin in the tubing and on the sucker rods, allowing it to be flushed out.

Another thing that can be done to remove paraffin is to scrape it out. For instance, on some pumping wells special *paraffin scrapers* are permanently attached to the sucker rods (fig. 112). As the rods move up and down in the tubing, the scrapers continuously remove any paraffin. Other types of scrapers can be used on flowing wells to remove paraffin.

Because of the crystalline structure of paraffin, deposits build up on rough surfaces such as the inside walls of tubing. Thus, coating the inside surfaces of tubing or flow lines with smooth plastic is effective in preventing the formation of paraffin.

Figure 112. Special paraffin scrapers are permanently attached to the sucker rods on some pumping wells.

Hydrogen Sulfide

Hydrogen sulfide, or H_2S, can be a serious hazard to people and equipment. It is a poisonous gas, heavier than air, flammable, and colorless. In small amounts, it smells similar to rotten eggs. H_2S deadens the sense of smell in large, or concentrated, amounts. And when H_2S is concentrated, it is very dangerous. Depending on how concentrated it is, H_2S can do everything from just smelling bad to irritating the eyes and skin. Or, it can kill very quickly.

In working on a lease that produces sour crude (oil with H_2S), special precautions must be taken. More than anything, it is a matter of being constantly aware that H_2S can be deadly. So, a worker does not stick his head into the thief hatch on a tank of sour crude. In fact, he does not go where there may be concentrations of H_2S without special protection. Protection from H_2S sometimes involves wearing breathing equipment similar to the scuba gear an underwater diver wears. Breathing equipment on a lease usually consists of an airtight mask and a supply of compressed air in a steel bottle carried on the user's back (fig. 113). Those who work on sour crude leases receive special training to show them how to safely work around H_2S. H_2S must be respected and good judgment used.

Figure 113. Protection from the harmful effects of H_2S includes the use of special breathing equipment.

Review

The purpose of this story is to introduce the reader to many of the processes, techniques, and equipment used in producing oil and gas. To review some of these, we recall that oil and gas do not occur in underground rivers or lakes. Instead, they occur in small, interconnected pore spaces in rock.

Oil and gas wells are usually drilled at places where seismic exploration techniques and geologic interpretation have indicated that hydrocarbons may exist in the subsurface. Drilling is really the only sure way to determine whether or not they actually reside there.

As a hole is drilled, it is lined with steel pipe, called casing, that is cemented into the hole. Usually, the final string of casing is cemented in place and perforated in order to complete the well. Sometimes, different completion schemes are utilized, depending on the nature of the reservoir and other factors.

Oil can flow out of a reservoir and up the well to the surface using natural gas and water energy that exists with the oil, or it may have to be lifted artificially. Two general methods of artificial lift are those that use pumps and those that use injected gas.

Even with artificial lift, all of the oil originally in the reservoir will not be recovered. In fact, three-fourths or more of it may remain in place, unless additional recovery techniques such as waterflooding and miscible processes are able to recover more. Even present-day additional recovery techniques will not recover all of the oil that is in a reservoir.

Whether oil flows to the surface through natural energy, artificial lift, or by way of additional recovery techniques, that oil is not pure. Crude oil from the reservoir usually has gas, water, and sediment in it. Most of the water and sediment have to be removed before the crude can be sold to a shipper. Also, gas and oil must be separated from each other. Special vessels called separators and treaters are employed to separate fluids and to remove emulsified water from oil.

Well tests are run on wells throughout their producing lives. Their basic purpose is to assist the well owner in producing his wells efficiently. Some of the tests measure the potential of the well; others measure bottomhole pressure and temperature; still others indicate the fluid level in a well. Regardless of the test, the ultimate goal is to recover as much oil from a reservoir as possible.

Oil must be temporarily stored on a lease before it is sold to a transportation company. While it is on the lease, oil is accurately tested and gauged, or measured, in order to determine its quality and quantity prior to sale.

Finally, some special problems may be encountered on the lease. Corrosion, scale, paraffin, H_2S, and saltwater disposal are some of the problems that must be solved if successful production is to continue. And production—producing oil and gas, the most popular form of energy ever used by mankind—is what this story is all about.

GLOSSARY

A

absorber *n:* 1. a vertical, cylindrical vessel that recovers heavier hydrocarbons from a mixture of predominantly lighter hydrocarbons. Also called absorption tower. 2. a vessel in which gas is dehydrated by being bubbled through glycol.

acetic acid *n:* an organic acid compound sometimes used to acidize oilwells. It is not as corrosive as other acids used in well treatments. Its formula is $C_2H_4O_2$, or CH_3COOH.

acidize *v:* to treat oil-bearing limestone or other formations with acid for the purpose of increasing production. Hydrochloric or other acid is injected into the formation under pressure. The acid etches the rock, enlarging the pore spaces and passages through which the reservoir fluids flow. The acid is held under pressure for a period of time and then pumped out, after which the well is swabbed and put back into production. Chemical inhibitors combined with the acid prevent corrosion of the pipe.

acoustic log *n:* a record of the measurement of porosity done by comparing depth to the time it takes for a sonic impulse to travel through a given length of formation. The rate of travel of the sound wave through a rock depends on the composition of the formation and the fluids it contains. Because the type of formation can be ascertained by other logs, and because sonic transit time varies with relative amounts of rock and fluid, porosity can usually be determined in this way.

acoustic well logging *n:* the process of recording the acoustic characteristics of subsurface formations, based on the time required for a sound wave to travel a specific distance through rock. The rate of travel of the sound wave depends on the composition of the formation, its porosity, and its fluid content.

air gun *n:* 1. a hand tool that is powered pneumatically. 2. a chamber filled with compressed air, often used offshore in seismic exploration. As the gun is trailed behind a boat, air is released, making a low-frequency popping noise, which penetrates the subsurface rock layers and is reflected by the layers. Sensitive hydrophones receive the reflections and transmit them to recording equipment on the boat.

alkali *n:* a substance having marked basic (alkaline) properties, such as a hydroxide of an alkali metal. See *base.*

American Petroleum Institute *n:* founded in 1920, this national oil trade organization is the leading standardizing organization for oil field drilling and producing equipment. It maintains departments of transportation, refining, and marketing in Washington, D.C., and a department of production in Dallas. *adj:* (slang) indicative of a job being properly or thoroughly done (as,"His work is strictly API").

annular space *n:* 1. the space surrounding a cylindrical object within a cylinder. 2. the space around a pipe in a wellbore, the outer wall of which may be the wall of either the borehole or the casing; sometimes termed the annulus.

annulus *n:* also called annular space. See *annular space.*

anticline *n:* an arched, inverted-trough configuration of folded and stratified rock layers. Compare *syncline.*

API *abbr:* American Petroleum Institute.

API gravity *n:* the measure of the density or gravity of liquid petroleum products in the United States, derived from specific gravity in accordance with the following equation:

$$\text{API gravity} = \frac{141.5}{\text{specific gravity}} - 131.5$$

API gravity is expressed in degrees, a specific gravity of 1.0 being equivalent to 10° API.

artificial lift *n:* any method used to raise oil to the surface through a well after reservoir pressure has declined to the point at which the well no longer produces by means of natural energy. Sucker rod pumps, gas lift, hydraulic pumps, and submersible electric pumps are the most common forms of artificial lift.

B

bad oil *n:* oil that contains a sufficient amount of basic sediment and water to make it unacceptable by a pipeline or other oil transportation system.

baffle plate *n:* 1. a partial restriction, generally a plate, placed to change the direction, guide the flow, or promote mixing within a tank or vessel. 2. a device that is seated on the bit pin, in a tool joint, or in a drill pipe float, used to centralize the lower end of a go-devil while permitting the bypass of drilling fluid. The go-devil contains a surveying instrument.

ball-and-seat valve *n:* a device used to restrict fluid flow to one direction, consisting of a polished sphere, or ball, usually of metal and an annular piece, the seat, ground and polished to form a seal with the surface of the ball. Gravitational force or the force of a spring holds the ball against the seat. Flow in the direction of the force is prevented, while flow in the opposite direction overcomes the force and unseats the ball.

barrel *n:* 1. a measure of volume for petroleum products in the United States. One barrel is the equivalent of 42 U.S. gallons or 0.15899 cubic metres. One cubic metre equals 6.2897 barrels. 2. the cylindrical part of a sucker rod pump in which the piston-like plunger moves up and down. Operating as a piston inside a cylinder, the plunger and barrel create pressure energy to lift well fluids to the surface.

base *n:* a substance capable of reacting with an acid to form a salt. A typical base is sodium hydroxide (caustic), with the chemical formula NaOH. For example, sodium hydroxide combines with hydrochloric acid to form sodium chloride (a salt) and water; this reaction is written chemically as

$$NaOH + HCl \longrightarrow NaCl + H_2O.$$

basic sediment and water *n:* the water and other extraneous material present in crude oil. Usually, the BS&W content must be quite low before a pipeline or other shipper will accept the oil for delivery to a refinery. The amount acceptable depends on a number of factors but usually runs from less than 5 percent to a small fraction of 1 percent.

beam pumping unit *n:* a machine designed specifically for sucker rod pumping, using a horizontal member (walking beam) that is worked up and down by a rotating crank to produce reciprocating motion.

bit *n:* the cutting or boring element used in drilling oil and gas wells. The bit consists of a cutting element and a circulating element. The circulating element permits the passage of drilling fluid and utilizes the hydraulic force of the fluid stream to improve drilling rates. In rotary drilling, several drill collars are joined to the bottom end of the drill pipe column, and the bit is attached to the end of the string of drill collars. Most bits used in rotary drilling are roller cone bits, but diamond bits are also used extensively.

blowout *n:* an uncontrolled flow of gas, oil, or other well fluids into the atmosphere. A blowout, or gusher, can occur when formation pressure exceeds the pressure applied to it by the column of drilling fluid. A kick warns of an impending blowout. See *gusher* and *kick.*

borehole *n:* the wellbore; the hole made by drilling or boring. See *wellbore.*

bottomhole pressure *n:* 1. the pressure at the bottom of a borehole. It is caused by the hydrostatic pressure of the drilling fluid in the hole and, sometimes, any back-pressure held at the surface, as when the well is shut in with blowout preventers. When mud is being circulated, bottomhole pressure is the hydrostatic pressure plus the remaining circulating pressure required to move the mud up the annulus. 2. the pressure in a well at a point opposite the producing formation, as recorded by a bottomhole pressure bomb.

bottomhole pressure gauge *n:* a gauge to measure bottomhole pressure. See *bottomhole pressure.*

bottomhole pressure test *n:* a test that measures the reservoir pressure of the well, obtained at a specific depth or at the midpoint of the producing zone. A flowing bottomhole pressure test measures pressure while the well continues to flow; a shut-in bottomhole pressure test measures pressure after the well has been shut in for a specified period of time. See *bottomhole pressure gauge* and *bottomhole pressure.*

bottomhole pump *n:* any of the rod pumps, high-pressure liquid pumps, or centrifugal pumps located at or near the bottom of the well and used to lift the well fluids. See *sucker rod pumping, hydraulic pumping,* and *submersible pump.*

bottomhole temperature *n:* temperature measured in a well at a depth at the midpoint of the thickness of the producing zone.

bridle *n:* a cable on a pumping unit, looped over the horsehead and connected to the carrier bar to support the polished-rod clamp. See *sucker rod pumping.*

bright spot *n:* a seismic phenomenon that shows up on a seismic, or record, section as a sound reflection that is much stronger than usual. A bright spot sometimes directly indicates natural gas in a trap. See *record section.*

brine *n:* water that has a large quantity of salt, especially sodium chloride, dissolved in it; salt water.

BS&W *abbr:* basic sediment and water.

bubble cap *n:* a metal cap, mounted on a tray, that has openings allowing vapor bubbles in a gas-processing tower to contact cool liquids, causing some of the vapor to condense to liquid.

bubble-cap tray *n:* a perforated steel tray on which bubble caps are mounted. Bubble caps and trays are arranged in bubble towers, cylindrical vessels set vertically.

C

cable *n:* a rope of wire, hemp, or other strong fibers. See *wire rope.*

calcium carbonate *n:* a chemical combination of calcium, carbon, and oxygen; the main constituent of limestone. It forms a tenacious scale in water-handling facilities and is a cause of water hardness. Chemical formula is $CaCO_3$.

calcium magnesium carbonate *n:* a chemical combination of calcium, magnesium, carbon, and oxygen; the main constituent of dolomite. Its chemical formula is $CaMgCO_3$. See *dolomite.* Compare *calcium carbonate.*

cap rock *n:* 1. impermeable rock overlying an oil or gas reservoir that tends to prevent migration of oil or gas out of the reservoir. 2. the porous and permeable strata overlying salt domes that may serve as the reservoir rock.

carbonate *n:* 1. a salt of carbonic acid. 2. a compound containing the carbonate (CO_3) radical.

carbon dioxide *n:* a colorless, odorless gaseous compound of carbon and oxygen (CO_2). A product of combustion and a filler for fire extinguishers, this heavier-than-air gas can collect in low-lying areas, where it may displace oxygen and present the hazard of anoxia.

casing *n:* steel pipe placed in an oil or gas well as drilling progresses to prevent the wall of the hole from caving in during drilling, to prevent seepage of fluids, and to provide a means of extracting petroleum if the well is productive.

casing hanger *n:* an arrangement of slips and packing rings, used to suspend casing from a casinghead.

casing-tubing annulus *n:* In a wellbore, the space between the inside of the casing and the outside of the tubing.

casinghead *n:* a heavy, flanged steel fitting connected to the first string of casing; it provides a housing for slips and packing assemblies, allows suspension of intermediate and production strings of casing, and supplies the means for the annulus to be sealed off. Also called a spool.

cathodic protection *n:* a method of protecting a metal structure from corrosion by making its surfaces cathodic and controlling the location of anodic areas so that corrosion damage can be reduced to tolerable levels.

caustic soda *n:* sodium hydroxide, used to maintain an alkaline pH in drilling mud and in petroleum fractions. Its formula is NaOH.

cement *n:* a powder, consisting of alumina, silica, lime, and other substances that hardens when mixed with water. Extensively used in the oil industry to bond casing to the walls of the wellbore.

cement channeling *n:* an undesirable phenomenon that can occur when casing is being cemented in a borehole. The cement slurry fails to rise uniformly between the casing and the borehole wall, leaving spaces devoid of cement. Ideally, the cement should completely and uniformly surround the casing and form a strong bond to the borehole wall.

channeling *n:* the bypassing of oil in a water-drive field due to erratic or uncontrolled water encroachment. The natural tendency toward channeling is aggravated by excessive production rates, which encourage premature water encroachment. See *cement channeling.*

chemical flooding *n:* See *micellar-polymer flooding.*

chemical pump *n:* an injection pump used to introduce a chemical into a fluid stream or receptacle.

choke *n:* a device with an orifice installed in a line to restrict the flow of fluids. Surface chokes are part of the Christmas tree on a well and contain a choke nipple, or bean, with a small-diameter bore that serves to restrict the flow. Chokes are also used to control the rate of flow of the drilling mud out of the hole when the well is closed in with the blowout preventer and a kick is being circulated out of the hole.

Christmas tree *n:* the control valves, pressure gauges, and chokes assembled at the top of a well to control the flow of oil and gas after the well has been drilled and completed.

circulation *n:* the movement of drilling fluid out of the mud pits, down the drill stem, up the annulus, and back to the mud pits.

coalesce *v:* to unite as a whole.

combination drive *n:* the natural energy that forces fluids from a reservoir and into a wellbore, provided by a gas cap above and water below the oil in the reservoir. See *reservoir drive mechanism, gas-cap drive,* and *water drive.*

complete a well *v:* to finish work on a well and bring it to productive status. See *well completion.*

condensation *n:* the process by which vapors are converted into liquids, chiefly accomplished by cooling the vapors. Condensation is often the cause of water appearing in fuels.

conductor casing *n:* generally, the first string of casing in a well. It may be lowered into a hole drilled into the formations near the surface and cemented in place, or it may be driven into the ground by a special pile driver. (In such cases, it is sometimes called drive pipe.) Its purpose is to prevent the soft formations near the surface from caving in and to conduct drilling mud from the bottom of the hole to the surface when drilling starts. Also called conductor pipe. See *conductor pipe.*

conductor pipe *n:* a short string of large-diameter casing used to keep the wellbore open and to provide a means of conveying the up-flowing drilling fluid from the wellbore to the mud pit.

connate water *n:* water retained in the pore spaces, or interstices, of a formation from the time the formation was created. Compare *interstitial water.*

core *n:* a cylindrical sample taken from a formation for geological analysis. Usually a conventional core barrel is substituted for the bit and procures a sample as it penetrates the formation. *v:* to obtain a formation sample for analysis.

corrosion *n:* any of a variety of complex chemical or electrochemical processes by which metal is destroyed through reaction with its environment. For example, rust is corrosion.

corrosion control *n:* the measures used to prevent or reduce the effects of corrosion. These practices can range from simply painting metal, to isolate it from moisture and chemicals and to insulate it from galvanic currents, to cathodic protection, in which a galvanic or impressed direct electric current renders a pipeline cathodic, thus causing it to be a negative element in the circuit. The use of chemical inhibitors and closed systems are other examples of corrosion control.

corrosion inhibitor *n:* a chemical substance that minimizes or prevents corrosion from occurring in metal equipment.

coupling *n:* 1. in piping, a metal collar with internal threads used to join two sections of threaded pipe. 2. in power transmission, a connection extending longitudinally between a driving shaft and a driven shaft. Most such couplings are flexible and compensate for minor misalignment of the two shafts.

crank arm *n:* a steel member connected to each end of the shaft extending from each side of the speed reducer on a beam pumping unit.

crude oil *n:* unrefined liquid petroleum. It ranges in gravity from 9° API to 55° API and in color from yellow to black, and it may have a paraffin, asphalt, or mixed base. If a crude oil, or crude, contains a sizable amount of sulfur or sulfur compounds, it is called a sour crude; if it has little or no sulfur, it is called a sweet crude. In addition, crude oils may be referred to as heavy or light according to API gravity, the lighter oils having the higher gravities.

cuttings *n pl:* the fragments of rock dislodged by the bit and brought to the surface in the drilling mud. They are not the same as cavings, which are particles that fall off the wall of the hole. Washed and dried samples of the cuttings are analyzed by geologists to obtain information about the formations drilled.

cyclic steam injection *n:* a method of treating heavy, viscous hydrocarbons so that they may be produced. Steam is injected in preset cycles into the reservoir, where it heats the hydrocarbons, making them less viscous and able to flow into the wellbore.

cylinder *n:* 1. the unit of an internal-combustion engine in which combustion and compression take place. 2. a chamber in a pump from which the piston expels fluid.

D

dehydration *n:* the removal of water or water vapor from gas or oil. See *dehydrator.*

dehydrator *n:* equipment or apparatus for effecting dehydration.

demulsifier *n:* a chemical with properties that cause the water droplets in a water-in-oil emulsion to merge and settle out of the oil when the chemical is added to the emulsion. Also called emulsion breaker. See *demulsify* and *emulsion breaker.*

demulsify *v:* to resolve an emulsion, especially of water and oil, into its components. See *emulsion treating.*

density *n:* the mass or weight of a substance per unit volume. For instance, the density of a drilling mud may be 10 pounds per gallon (ppg), 74.8 pounds per cubic foot (lb/ft^3), or 198.2 kilograms per cubic metre (kg/m^3). Specific gravity, relative density, and API gravity are other units of density. See *API gravity* and *specific gravity.*

deplete *v:* to exhaust a supply. An oil and gas reservoir is depleted when most or all economically recoverable hydrocarbons have been produced.

dissolved-gas drive *n:* a solution-gas drive. See *reservoir drive mechanism.*

dolomite *n:* a type of sedimentary rock similar to limestone but rich in magnesium carbonate; sometimes a reservoir rock for petroleum.

dome *n:* a geologic structure resembling an inverted bowl; a short anticline that plunges on all sides.

double-box coupling *n:* a special coupling that has a female connection (box) on each end and that is used to connect sucker rods that have a male connection (pin) on each end.

downcomer *n:* a tube that conducts liquids downward in a vessel such as an absorber.

drill *v:* to bore a hole in the earth, usually to find and remove subsurface formation fluids such as oil and gas.

drilling fluid *n:* circulating fluid, one function of which is to force cuttings out of the wellbore and to the surface. Other functions are to cool the bit and to counteract downhole formation pressure. While a mixture of barite, clay, water, and chemical additives is the most common drilling fluid, wells can also be drilled by using air, gas, water, or oil-base mud as the drilling fluid. See *mud.*

drill pipe *n:* heavy seamless tubing used to rotate the bit and circulate the drilling fluid. Joints of pipe approximately 30 feet (9 m) long are coupled together by means of tool joints.

drill stem test *n:* the conventional method of formation testing. The basic drill stem test tool consists of a packer or packers, valves or ports that may be opened and closed from the surface, and two or more pressure-recording devices. The tool is lowered on the drill string to the zone to be tested. The packer or packers are set to isolate the zone from the drilling fluid column. The valves or ports are then opened, to allow for formation flow while the recorders chart flow pressures, and are then closed, to shut in the formation while the recorders chart static pressures. A sampling chamber traps clean formation fluids at the end of the test. Analysis of the pressure charts is an important part of formation testing.

dry hole *n:* any well that does not produce oil or gas in commercial quantities. A dry hole may flow water, gas, or even oil, but not enough to justify production.

duster *n:* a dry hole.

E

electric log *n:* also called an electric well log. See *electric well log.*

electric well log *n:* a record of certain electrical characteristics of formations traversed by the borehole, made to identify the formations, determine the nature and amount of fluids they contain, and estimate their depth. Also called an electric log or electric survey.

electrostatic treater *n:* a vessel that receives emulsion and resolves the emulsion to oil, water, and usually gas, by using heat, chemicals, and a high-voltage electric field. This field, produced by grids placed perpendicular to the flow of fluids in the treater, aids in breaking the emulsion. Also called an electrochemical treater. See *emulsion treating.*

emulsion *n:* a mixture in which one liquid, termed the dispersed phase, is uniformly distributed (usually as minute globules) in another liquid, called the continuous phase or dispersion medium. In an oil-water emulsion, the oil is the dispersed phase and the water the dispersion medium; in a water-oil emulsion, the reverse holds. A typical product of oilwells, water-oil emulsion is also used as a drilling fluid.

emulsion breaker *n:* a system, device, or process used for breaking down an emulsion and producing two or more easily separated compounds (such as water and oil). Emulsion breakers may be (1) devices to heat the emulsion, thus achieving separation by lowering the viscosity of the emulsion and allowing the water to settle out; (2) chemical compounds, which destroy or weaken the film around each globule of water, thus uniting all the drops; (3) mechanical devices such as settling tanks and wash tanks; or (4) electrostatic treaters, which use an electric field to cause coalescence of the water globules.

emulsion treating *n:* the process of breaking down emulsions to separate oil from water or other contaminants. Treating plants may use a single process or a combination of processes to effect demulsification, depending upon what emulsion is being treated.

engine *n:* a machine for converting the heat content of fuel into rotary motion that can be used to power other machines. An engine is not a motor.

enhanced oil recovery *n:* 1. the introduction of an artificial drive and displacement mechanism into a reservoir to produce oil unrecoverable by primary recovery methods. The purpose of enhanced oil recovery (EOR) is to restore formation pressure and fluid flow to a substantial portion of a reservoir by injecting fluids into injection wells located in rock that has fluid communication with production wells. EOR methods include waterflooding, chemical flooding, gas injection, and thermal recovery. Chemical flooding, most types of gas injection, and thermal methods are often called advanced EOR methods because they not only restore formation pressure but also improve displacement of oil by overcoming forces that keep the oil trapped in rock pores. 2. the use of an advanced EOR method. See *primary recovery, secondary recovery, tertiary recovery,* and the various EOR methods.

EOR *abbr:* enhanced oil recovery.

erosion *n:* the process by which material (such as rock or soil) is worn away or removed (as by wind or water).

exchanger *n:* a piping arrangement that permits heat from one fluid to be transferred to another fluid as they travel countercurrent to one another. In the heat exchanger of an emulsion-treating unit, heat from the outgoing clean oil is transferred to the incoming well fluid, cooling the oil and heating the well fluid.

exploration *n:* the search for reservoirs of oil and gas, including aerial and geophysical surveys, geological studies, core testing, and drilling of wildcats.

F

fault *n:* a break in subsurface strata. Often strata on one side of the fault line have been displaced (upward, downward, or laterally) relative to their original positions.

fault trap *n:* a subsurface hydrocarbon trap created by faulting, which causes an impermeable rock layer to be moved opposite the reservoir bed.

filter *n:* a porous medium through which a fluid is passed to separate particles of suspended solids from it.

fingering *n:* a phenomenon that often occurs in an injection well in which the fluid being injected does not contact the entire reservoir but bypasses sections of the reservoir fluids in a fingerlike manner. Fingering is not desirable because portions of the reservoir are not contacted by the injection fluid.

fire flood *n:* also called *in situ* combustion. See *in situ combustion.*

fire tube *n:* A pipe, or set of pipes, within an emulsion settling tank through which steam or hot gases are passed to warm the emulsion and make the oil less viscous.

fissure *n:* a crack or fracture in a subsurface formation.

flow *n:* a current or stream of fluid.

flow line *n:* the surface pipe through which oil travels from a well to processing equipment or to storage.

flowing bottomhole pressure test *n:* a bottomhole pressure test that measures pressure while the well is flowing. See *bottomhole pressure test.*

fluid *n:* a substance that flows and yields to any force tending to change its shape. Liquids and gases are fluids.

fold *n:* a flexure of rock strata into arches and troughs, produced by earth movements. See *anticline* and *syncline.*

formation *n:* a bed or deposit composed throughout of substantially the same kind of rock; a lithologic unit. Each different formation is given a name, frequently as a result of the study of the formation outcrop at the surface and sometimes based on fossils found in the formation.

formation damage *n:* the reduction of permeability in a reservoir rock caused by the invasion of drilling fluid and treating fluids to the section adjacent to the wellbore. It is often called skin.

formation fracturing *n:* a method of stimulating production by increasing the permeability of the producing formation. Under extremely high hydraulic pressure, a fluid (such as water, oil, alcohol, dilute hydrocholoric acid, liquefied petroleum gas, or foam) is pumped downward through tubing or drill pipe and forced into the perforations in the casing. The fluid enters the formation and parts or fractures it. Sand grains, aluminum pellets, glass beads, or similar materials are carried in suspension by the fluid into the fractures. These are called propping agents or proppants. When the pressure is released at the surface, the fracturing fluid returns to the well, and the fractures partially close on the proppants, leaving channels for oil to flow through them to the well. This process is often called a frac job.

formic acid *n:* a simple organic acid used for acidizing oilwells. It is stronger than acetic acid but much less corrosive than hydrofluoric or hydrochloric acid and is usually used for high-temperature wells.

frac job *n:* See *formation fracturing.*

fracture *n:* a crack or crevice in a formation, either natural or induced.

fracture acidizing *n:* a procedure by which acid is forced into a formation under pressure high enough to cause the formation to crack. The acid acts on certain kinds of rocks, usually carbonates, to increase the permeability of the formation. Compare *matrix acidizing.*

fracturing *n:* shortened form of formation fracturing. See *formation fracturing.*

free water *n:* 1. the water produced with oil. It usually settles out within five minutes when the well fluids become stationary in a settling space within a vessel. 2. the measured volume of water that is present in a container and that is not in suspension in the contained liquid at observed temperature.

G

gas *n:* a compressible fluid that completely fills any container in which it is confined. Technically, a gas will not condense when it is compressed and cooled, because a gas can exist only above the critical temperature for its particular composition. Below the critical temperature, this form of matter is known as a vapor because liquid can exist and condensation can occur. Sometimes the terms gas and vapor are used interchangeably. However, the term *vapor* should be used for those streams in which condensation can occur and which originate from, or are in equilibrium with, a liquid phase.

gas cap *n:* a free-gas phase overlying an oil zone and occurring within the same producing formation as the oil. See *reservoir.*

gas-cap drive *n:* drive energy supplied naturally (as a reservoir is produced) by the expansion of gas in a cap overlying the oil in the reservoir. See *reservoir drive mechanism.*

gas drive *n:* the use of the energy that arises from gas compressed in a reservoir to move crude oil to a wellbore. Gas drive is also used in a form of secondary recovery in which gas is injected into wells to sweep remaining oil to a producing well.

gas injection *n:* 1. the injection into a reservoir of natural gas produced from the reservoir in order to maintain formation pressure and reduce the rate of decline of the original reservoir drive. Used for pressure maintenance, this method is called gas recycling or immiscible gas injection. 2. injection of a gas into a reservoir to restore reservoir pressure and to free trapped oil by miscible displacement. Used for enhanced oil recovery, this method is called miscible gas injection. A variety of gases may be used, including liquid petroleum gas, methane enriched with other light hydrocarbons, methane under high pressure, nitrogen under high pressure, and carbon dioxide. Frequently, water is also injected in alternating slugs with the gas.

gas lift *n:* the process of raising or lifting fluid from a well by injecting gas down the well through tubing or through the tubing-casing annulus. Injected gas aerates the fluid to make it exert less pressure than the formation does; consequently, the higher formation pressure forces the fluid out of the wellbore. Gas may be injected continuously or intermittently, depending on the producing characteristics of the well and the arrangement of the gas-lift equipment.

gas-lift valve *n:* a device installed on a gas-lift mandrel, which in turn is put on the tubing string of a gas-lift well. Tubing and casing pressures cause the valve to open and close, thus allowing gas to be injected into the fluid in the tubing to cause the fluid to rise to the surface.

gas liquids *n pl:* See *liquefied petroleum gas.*

gas well *n:* a well that primarily produces gas. Legal definitions vary among the states.

gauge *n:* 1. the diameter of a bit or the hole drilled by the bit. 2. a device used to measure some physical property (such as a pressure gauge). *v:* to measure size, volume, or other measurable property.

gauging tape *n:* a metal tape used to measure the depth of liquid in a tank.

geologist *n:* a scientist who gathers and interprets data pertaining to the strata of the earth's crust.

geophone *n:* an instrument that detects vibrations passing through the earth's crust, used in conjunction with seismography. Geophones are often called jugs. See *seismograph.*

geophysicist *n:* one who studies the physics of the earth, including meteorology, hydrology, oceanography, seismology, volcanology, magnetism, and radioactivity.

glycol *n:* a group of compounds used to dehydrate gaseous or liquid hydrocarbons or to inhibit the formation of hydrates. Commonly used glycols are ethylene glycol, diethylene glycol, and triethylene glycol.

glycol absorber *n:* a cylinder consisting of several perforated trays with bubble caps mounted over the perforations and functioning as a part of a glycol dehydration system.

glycol dehydrator *n:* a processing unit used to remove all or most of the water from gas. Usually a glycol unit includes an absorber, in which the wet gas is put into contact with glycol to remove the water, and a reboiler, which heats the wet glycol to remove the water from it so that it can be recycled.

gravel-pack *v:* to place a slotted or perforated liner in a well and surround it with small-sized gravel.

gravel packing *n:* a method of well completion in which a slotted or perforated liner is placed in the well and surrounded by small-sized gravel. The well is sometimes enlarged by underreaming at the point where the gravel is packed. The mass of gravel excludes sand from intruding in the well but allows continued rapid production.

gravity *n:* the attraction exerted by the earth's mass on objects at its surface; the weight of a body. See *API gravity* and *specific gravity.*

gusher *n:* an oilwell that has come in with such great pressure that the oil jets out of the well like a geyser. In reality, a gusher is a blowout and is extremely wasteful of reservoir fluids and drive energy. In the early days of the oil industry, gushers were common and many times were the only indications that a large reservoir of oil and gas had been struck. See *blowout.*

H

hanger *n:* See *casing hanger* and *tubing hanger.*

hanger-packer *n:* a device that fits around the top of a string of casing or a string of tubing and supports part of the weight of the casing or tubing while providing a pressure-tight seal. See *hanger* and *packer.*

heater-treater *n:* a vessel that heats an emulsion and removes water and gas from the oil to raise it to a quality acceptable for pipeline transmission. A heater-treater is a combination of heater, free-water knockout, and oil and gas separator.

heat exchanger *n:* See *exchanger.*

homogeneous rock *n:* rock whose characteristics (permeability, porosity, density, etc.) are generally similar throughout.

horizontal heater-treater *n:* a heater-treater of cylindrical shape lying parallel to the ground. See *heater-treater.*

horizontal separator *n:* a separator of cylindrical shape lying parallel to the ground. See *separator.*

horsehead *n:* the curved section of the walking beam of a beam pumping unit, which is located on the oilwell end and from which the bridle is suspended.

huff-and-puff injection *n:* (slang) cyclic steam injection.

hydrate *n:* a hydrocarbon and water compound that is formed under reduced temperature and pressure in gathering, compression, and transmission facilities for gas. Hydrates often accumulate in troublesome amounts and impede fluid flow. They resemble snow or ice.

hydraulic *adj:* 1. of or relating to water or other liquid in motion. 2. operated, moved, or effected by water or liquid.

hydraulic fracturing *n:* an operation in which a specially blended liquid is pumped down a well and into a formation under pressure high enough to cause the formation to crack open. The resulting cracks or fractures serve as passages through which oil can flow into the wellbore. See *formation fracturing.*

hydraulic pump *n:* a device that lifts oil from wells without the use of sucker rods. See *hydraulic pumping.*

hydraulic pumping *n:* a method of pumping oil from wells by using a downhole pump without sucker rods. Subsurface hydraulic pumps consist of two reciprocating pumps coupled and placed in the well. One pump functions as an engine and drives the other pump (the production pump). Surface power is supplied from a standard engine-driven pump. The downhole engine is operated by clean crude oil under pressure (power oil) that is drawn from a power-oil or settling tank by a triplex plunger pump. If a single string of tubing is used, power oil is pumped down the tubing string to the pump, which is seated in the string, and a mixture of power oil and produced fluid is returned through the casing-tubing annulus. If two parallel strings are used, one supplies the power oil to the pump while the other returns the exhaust and the produced oil to the surface. The hydraulic pump may be used to pump several wells from a central source and has been used to lift oil from depths of more than 10,000 feet (3 048 m).

hydraulics *n:* the branch of science that deals with practical applications of water or other liquid in motion.

hydrocarbons *n pl:* organic compounds of hydrogen and carbon, whose densities, boiling points, and freezing points increase as their molecular weights increase. Although composed of only two elements, hydrocarbons exist in a variety of compounds because of the strong affinity of the carbon atom for other atoms and for itself. The smallest molecules of hydrocarbons are gaseous; the largest are solids. Petroleum is a mixture of many different hydrocarbons.

hydrochloric acid *n:* an acid compound commonly used to acidize carbonate rocks; prepared by mixing hydrogen chloride gas in water. Also known as muriatic acid. Its chemical formula is HCl.

hydrofluoric acid *n:* a strong, poisonous liquid acid compound of hydrogen and fluorine (HF). Often mixed with hydrochloric acid, it is used mainly to remove mud from the wellbore and surrounding formation pores. Also called mud acid.

hydrogen sulfide *n:* a flammable, colorless gaseous compound of hydrogen and sulfur (H_2S) with the odor of rotten eggs. Commonly found in petroleum, it causes the foul smell of petroleum fractions. It is extremely corrosive and poisonous, causing damage to skin, eyes, breathing passages, and lungs and attacking and paralyzing the nervous system, particularly that part controlling the lungs and heart. Also called hepatic gas or sulfureted hydrogen.

hydrophone *n:* a device, trailed in an array behind a boat in offshore seismic exploration, that is used to detect sound reflections, convert them to electric current, and send them through a cable to recording equipment on the boat.

H_2S *form:* hydrogen sulfide.

I

impeller *n:* a set of mounted blades used to impart motion to a fluid (e.g., the rotor of a centrifugal pump).

indirect heater *n:* apparatus or equipment in which heat from a primary source, usually the combustion of fuel, is transferred to a fluid or solid which acts as the heating medium.

insert pump *n:* a sucker rod pump that is run into the well as a complete unit. See *sucker rod pump.*

in situ combustion *n:* a method of enhanced oil recovery in which heat is generated within the reservoir by injecting air and burning a portion of the oil in place. The heat of initial combustion cracks the crude hydrocarbons, vaporizes the lighter hydrocarbons, and deposits the heavier hydrocarbons as coke. As the fire moves from the injection well in the direction of producing wells, it burns the deposited coke, releases hot combustion gases, and converts connate water into steam. The vaporized hydrocarbons and the steam move ahead of the combustion zone, condensing into liquids as they cool and moving oil by miscible displacement and hot waterflooding. Combustion gases provide additional gas drive. Heat lowers the viscosity of the oil, causing it to flow more freely. This method is used to recover heavy, viscous oil. Also called fire flooding.

intermediate casing string *n:* the string of casing set in a well after the surface casing is set to keep the hole from caving and to seal off troublesome formations. The string is sometimes called protection casing.

interstitial water *n:* water contained in the interstices of reservoir rock. In reservoir engineering, it is synonymous with connate water. See *connate water.*

J

jet *n:* 1. a hydraulic device operated by pump pressure to clean mud pits and tanks in rotary drilling and to mix mud components. 2. in a perforating gun using shaped charges, a highly penetrating, fast-moving stream of exploded particles that cuts a hole in the casing, cement, and formation.

joint *n:* a single length (30 feet or 9 m) of drill pipe, drill collar, casing, or tubing that has threaded connections at both ends. Several joints screwed together constitute a stand of pipe.

K

kick *n:* an entry of water, gas, oil, or other formation fluid into the wellbore during drilling. It occurs because the pressure exerted by the column of drilling fluid is not great enough to overcome the pressure exerted by the fluids in the formation drilled. If prompt action is not taken to control the kick or kill the well, a blowout may occur.

L

LACT unit *n:* an automated system for measuring, testing, and transferring oil from a lease gathering system into a pipeline. See *lease automatic custody transfer.*

lease *n:* 1. a legal document executed between a landowner, as lessor, and a company or individual, as lessee, that grants the right to exploit the premises for minerals or other products. 2. the area where production wells, stock tanks, separators, LACT units, and other production equipment are located.

lease automatic custody transfer *n:* the measurement, testing, and transfer of oil from the producer's tanks to the connected pipeline on an automatic basis without a representative of either the producer or the gathering company having to be present. See *LACT unit.*

limestone *n:* a sedimentary rock rich in calcium carbonate; sometimes serves as a reservoir rock for petroleum.

liner *n:* 1. a string of pipe used to case open hole below existing casing. A liner extends from the setting depth up into another string of casing, usually overlapping about 100 feet above the lower end of the intermediate or oil string. Liners are nearly always suspended from the upper string by a hanger device. 2. in jet perforating guns, a conically shaped metallic piece that is part of a shaped charge. It increases the efficiency of the charge by increasing the penetrating ability of the jet. 3. a replaceable tube that fits inside the cylinder of an engine or a pump.

liquid *n:* a state of matter in which the shape of the given mass depends on the containing vessel, but the volume of the mass is independent of the vessel. A liquid is a fluid that is almost incompressible.

liquid-level gauge *n:* any device that indicates the level or quantity of liquid in a container.

liquefied petroleum gas *n:* a mixture of heavier, gaseous, paraffinic hydrocarbons, principally butane and propane. These gases, easily liquefied at moderate pressure, may be transported as liquids but converted to gases on release of the pressure. Thus, liquefied petroleum gas is a portable source of thermal energy that finds wide application in areas where it is impractical to distribute natural gas. It is also used as a fuel for internal-combustion engines and has many industrial and domestic uses. Principal sources are natural and refinery gas, from which the liquefied petroleum gases are separated by fractionation.

log *n:* a systematic recording of data, such as a driller's log, mud log, electrical well log, or radioactivity log. Many different logs are run in wells to obtain various characteristics of downhole formations. *v:* to record data.

lubricator *n:* a specially fabricated length of casing or tubing usually placed temporarily above a valve on top of the casinghead or tubing head; used to run swabbing or perforating tools into a producing well; provides a method for sealing off pressure and thus should be rated for highest anticipated pressure.

M

master valve *n:* 1. a large valve located on the Christmas tree and used to control the flow of oil and gas from a well. Also called a master gate. 2. the blind or blank rams of a blowout preventer.

matrix acidizing *n:* the procedure by which acid flow is confined to the natural permeability and porosity of the formation, as opposed to fracture acidizing.

meter *n:* a device used to measure and often record volumes, quantities, or rates of flow of gases, liquids, or electric currents. *v:* to measure quantities or properties of a substance.

metering separator *n:* a complete separator and volume meter integrated into a single vessel. Two-phase units separate oil and gas and meter the oil; three-phase units separate oil, water, and gas, and meter the oil and water.

micellar-polymer flooding *n:* a method of enhanced oil recovery in which chemicals dissolved in water are pumped into a reservoir through injection wells in order to mobilize oil left behind after primary recovery and to move it toward producing wells. The chemical solution includes surfactants or surfactant-forming chemicals that reduce the interfacial and capillary forces between oil and water, dissolve the oil, and carry it out of the pores where it has been trapped. The solution may also contain cosurfactants to match the viscosity of the solution to that of the oil, to stabilize the solution, and to prevent its adsorption by reservoir rock. An electrolyte is often added to aid in adjusting viscosity. Injection of the chemical solution is followed by a slug of water thickened with a polymer, which pushes the dissolved oil through the reservoir, decreases the permeability of established channels so that new channels are opened, and serves as a mobility buffer between the chemical solution and the final injection of water.

miscible *adj:* capable of being mixed; capable of mixing in any ratio without separation of the two phases.

miscible drive *n:* a method of enhanced recovery in which various hydrocarbon solvents or gases (such as propane, LPG, natural gas, carbon dioxide, or a mixture thereof) are injected into the reservoir to reduce interfacial forces between oil and water in the pore channels and thus displace oil from the reservoir rock.

miscible flood *n:* a method of secondary recovery of fluids from a reservoir by injection of fluids that are miscible with the reservoir fluids.

mist extractor *n:* a metal member used to remove small droplets of moisture or condensable hydrocarbons from a gas stream in an oil and gas separator. The small droplets of moisture collect on the metal surface to form larger drops, which are removed from the separator along with other separated liquids.

mud *n:* the liquid circulated through the wellbore during rotary drilling and workover operations. In addition to its function of bringing cuttings to the surface, drilling mud cools and lubricates the bit and drill stem, protects against blowouts by holding back subsurface pressures, and deposits a mud cake on the wall of the borehole to prevent loss of fluids to the formation. Although it was originally a suspension of earth solids (especially clays) in water, the mud used in modern drilling operations is a more complex, three-phase mixture of liquids, reactive solids, and inert solids. The liquid phase may be fresh water, diesel oil, or crude oil and may contain one or more conditioners. See *drilling fluid.*

mud acid *n:* a mixture of hydrochloric and hydrofluoric acids and surfactants, used to effect mud removal from the wellbore.

mud solids *n pl:* the solid components of drilling mud. They may be added intentionally (barite, for example), or they may be introduced into the mud from the formation as the bit drills ahead. Usually used to refer to the latter.

multiple completion *n:* an arrangement for producing a well in which one wellbore penetrates two or more petroleum-bearing formations that lie one over the other. The tubing strings are suspended side by side in the production casing string, each a different length and each packed off to prevent the commingling of different reservoir fluids. Each reservoir is then produced through its own tubing string. Alternately, a small-diameter production casing string may be provided for each reservoir, as in multiple miniaturized or multiple tubingless completion.

N

natural drive energy *n:* See *reservoir drive mechanism.*

natural gas *n:* a highly compressible, highly expansible mixture of hydrocarbons having a low specific gravity and occurring naturally in gaseous form. Besides hydrocarbon gases, natural gas may contain appreciable quantities of nitrogen, helium, carbon dioxide, hydrogen sulfide, and water vapor. Although gaseous at normal temperatures and pressures, the gases comprising the mixture that is natural gas are variable in form and may be found either as gases or as liquids under suitable conditions of temperature and pressure.

net-oil computer *n:* a system of electronic and mechanical devices that automatically determines the amount of oil in a water and oil emulsion. One advantage of a net-oil computer is that the water and oil do not have to be separated for measuring the volume of the oil.

O

oil *n:* a simple or complex liquid mixture of hydrocarbons, which can be refined to yield gasoline, kerosine, diesel fuel, and various other products.

oil and gas separator *n:* an item of production equipment used to separate liquid components of the well stream from gaseous elements. Separators are either vertical or horizontal and either cylindrical or spherical in shape. Separation is accomplished principally by gravity, the heavier liquids falling to the bottom and the gas rising to the top. A float valve or other liquid-level control regulates the level of oil in the bottom of the separator.

oil field *n:* the surface area overlying an oil reservoir or reservoirs. Commonly, the term includes not only the surface area, but also the reservoir, the wells, and the production equipment.

oil field emulsion *n:* a combination of oil and water in which droplets of water are dispersed and suspended in the oil or droplets of oil are dispersed and suspended in the water. Most oil field emulsions are water-in-oil emulsions; that is, the water is dispersed in the oil. In order to remove the water from the oil, the emulsion must be treated with heat, chemicals, and/or electricity.

oil in place *n:* crude oil that is estimated to exist in a reservoir but has not been produced.

oil patch *n:* (slang) the oil field.

oil pool *n:* a loose term for an underground reservoir where oil is stored. Actually, oil is found in the pores of rocks, not in a pool.

oil sales outlet *n:* a drain located about a foot from the bottom of a stock tank, used to take oil out of the tank for sale to a pipeline company or other shipper. The space below the outlet allows room for any remaining basic sediment and water to settle out of the oil. Also called sales outlet and oil outlet.

oil shale *n:* a formation containing hydrocarbons that cannot be recovered by an ordinary oilwell but can be mined. After processing, the hydrocarbons are extracted from the shale. The cost of mining and treatment of oil shale has until recently been too great to compete with the cost of oilwell drilling.

oilwell *n:* a well from which oil is obtained.

oil zone *n:* a formation or horizon of a well from which oil may be produced. The oil zone is usually immediately under the gas zone and on top of the water zone if all three fluids are present and segregated.

open hole *n:* 1. any wellbore in which casing has not been set. 2. open or cased hole in which no drill pipe or tubing is suspended. 3. the portion of the wellbore that has no casing.

open-hole completion *n:* a method of preparing a well for production in which no production casing or liner is set opposite the producing formation. Reservoir fluids flow unrestricted into the open wellbore. An open-hole completion has limited use in rather special situations. Also called a barefoot completion.

organic theory *n:* an explanation of the origin of petroleum, which holds that the hydrogen and the carbon that make up petroleum come from plants and animals of land and sea. Furthermore, the theory holds that more of this organic material comes from very tiny creatures of swamp and sea than comes from larger creatures of land.

P

packer *n:* a piece of downhole equipment, consisting of a sealing device, a holding or setting device, and an inside passage for fluids, used to block the flow of fluids through the annular space between the tubing and the wall of the wellbore by sealing off the space between them. It is usually made up in the tubing string some distance above the producing zone. A packing element expands to prevent fluid flow except through the inside bore of the packer and into the tubing. Packers are classified according to configuration, use, and method of setting and whether or not they are retrievable (that is, whether they can be removed when necessary, or whether they must be milled or drilled out and thus destroyed).

paraffin *n:* a hydrocarbon having the formula C_nH_{2n+2} (e.g., methane, CH_4; ethane, C_2H_6). Heavier paraffin hydrocarbons (i.e., $C_{18}H_{38}$ and heavier) form a waxlike substance that is called paraffin. These heavier paraffins often accumulate on the walls of tubing and other production equipment, restricting or stopping the flow of desirable lighter paraffins.

paraffin scraper *n:* any tool used to remove paraffin from inside tubular goods.

pay sand *n:* the producing formation, often one that is not even sandstone. It is also called pay, pay zone, and producing zone.

perforate *v:* to pierce the casing wall and cement to provide holes through which formation fluids may enter or to provide holes in the casing so that materials may be introduced into the annulus between the casing and the wall of the borehole. Perforating is accomplished by lowering into the well a perforating gun, or perforator, that fires electrically detonated bullets or shaped charges.

perforated completion *n:* a well completion in which the production casing or liner is punctured to allow passage between the wellbore and the producing formation. Perforations are usually made with bullet- or jet-perforating guns.

perforated liner *n:* a liner that has had holes shot in it by a perforating gun. See *liner.*

perforating gun *n:* a device, fitted with shaped charges or bullets, that is lowered to the desired depth in a well and fired to create penetrating holes in casing, cement, and formation.

perforation *n:* a hole made in the casing, cement, and formation, through which formation fluids enter a wellbore. Usually several perforations are made at a time.

permeability *n:* 1. a measure of the ease with which fluids can flow through a porous rock. 2. the fluid conductivity of a porous medium. 3. the ability of a fluid to flow within the interconnected pore network of a porous medium.

petroleum *n:* a substance occurring naturally in the earth and composed mainly of mixtures of chemical compounds of carbon and hydrogen, with or without other nonmetallic elements such as sulfur, oxygen, and nitrogen. The compounds that compose it may be in the gaseous, liquid, or solid state, depending on their nature and on the existent conditions of temperature and pressure.

pin *n:* 1. the male section of a tool joint. 2. on a bit, the bit shank. 3. one of the pegs that are fitted on each side into the link plates (side bars) of a chain link of roller chain and that serve as the stable members onto which bushings are press-fitted and around which rollers move.

pipeline *n:* a system of connected lengths of pipe, usually buried in the earth or laid on the seafloor, that is used for transporting petroleum and natural gas.

piston *n:* a cylindrical sliding piece that is moved by or moves against fluid pressure within a confining cylindrical vessel.

pitman *n:* the arm that connects the crank to the walking beam on a pumping unit by means of which rotary motion is converted to reciprocating motion.

plug *n:* any object or device that blocks a hole or passageway (as a cement plug in a borehole).

plug and abandon *v:* to place cement plugs into a dry hole and abandon it.

plug trap *n:* a geologic configuration in which an intrusive body of rock has penetrated through and deformed one or more surrounding rock layers in such a way that conditions exist for hydrocarbons to accumulate.

plunger *n:* 1. a basic component of the sucker rod pump. See *sucker rod pump.* 2. the rod that serves as a piston in a reciprocating pump.

polished rod *n:* the topmost portion of a string of sucker rods, used for lifting fluid by the rod-pumping method. It has a uniform diameter and is smoothly polished to effectively seal pressure in the stuffing box attached to the top of the well.

polymer *n:* a substance that consists of large molecules formed from smaller molecules in repeating structural units. In petroleum refining, heat and pressure are used to polymerize light hydrocarbons into larger molecules, such as those that make up high-octane gasoline. In oilfield operations, various types of organic polymers are used to thicken drilling mud, fracturing fluid, acid, and other liquids. In petrochemical production, polymer hydrocarbons are used as the basis for plastics.

pore *n:* an opening or space within a rock or mass of rocks, usually small and often filled with some fluid (water, oil, gas, or all three). Compare *vug.*

porosity *n:* the condition of something that contains pores (such as a rock formation).

porous *adj:* having pores, or tiny openings, as in rock.

potential *n:* the maximum volume of oil or gas that a well is capable of producing.

potential test *n:* a test of the rate at which a well can produce oil or gas. See *potential.*

power oil *n:* in subsurface hydraulic pumping, the crude oil that is produced by a well, cleaned, and pumped into the well to energize the subsurface pump.

pressure *n:* the force that a fluid (liquid or gas) exerts uniformly in all directions within a vessel, pipe, hole in the ground, and so forth, such as that exerted against the liner wall of a tank or that exerted on the bottom of the wellbore by drilling mud. Pressure is expressed in terms of force exerted per unit of area, as pounds per square inch (psi) or grams or kilograms per square centimetre.

pressure maintenance *n:* the use of waterflooding or natural gas recycling during primary recovery to provide additional formation pressure and displacement energy that can supplement and conserve natural reservoir drives. Although commonly begun during primary production, pressure maintenance methods are often considered to be a form of enhanced oil recovery. See *enhanced oil recovery* and *secondary recovery.*

primary recovery *n:* the first stage of oil production in which natural reservoir drives are used to recover oil. When natural drives are still operative but incapable of producing profitable amounts of oil, they may be boosted by various types of artificial lift, well stimulation, and pressure maintenance. See *artificial lift, well stimulation,* and *pressure maintenance.*

prime mover *n:* an internal-combustion engine or a turbine that is the source of power for driving a machine or machines.

producer *n:* 1. a well that produces oil or gas in commercial quantities. 2. an operating company or individual in the business of producing oil; commonly called the operator.

producing zone *n:* the zone or formation from which oil or gas is produced. See *pay sand.*

production *n:* 1. the phase of the petroleum industry that deals with bringing the well fluids to the surface and separating them and with storing, gauging, and otherwise preparing the product for the pipeline. 2. the amount of oil or gas produced in a given period.

production casing *n:* the last string of casing that is set in a well, inside of which is usually suspended a tubing string.

production tank *n:* a tank used in the field to receive crude oil as it comes from the well. Also called a flow tank or lease tank.

productivity test *n:* a combination of a potential test and a bottomhole pressure test; its purpose is to determine the effects of different flow rates on the pressure within the producing zone of the well in order to establish physical characteristics of the reservoir and to determine the maximum potential rate of flow. See *potential test* and *bottomhole pressure test.*

proppant *n:* also called propping agent. See *propping agent.*

propping agent *n:* a granular substance (sand grains, aluminum pellets, or other material) that is carried in suspension by the fracturing fluid and that serves to keep the cracks open when fracturing fluid is withdrawn after a fracture treatment.

R

radioactivity log *n:* a record of the natural or induced radioactive characteristics of subsurface formations. See *radioactivity well logging.*

radioactivity well logging *n:* the recording of the natural or induced radioactive characteristics of subsurface formations. A radioactivity log, also known as a radiation log, normally consists of two recorded curves: a gamma ray curve and a neutron curve. Both help to determine the types of rocks in the formation and the types of fluids contained in the rocks. The two logs may be run simultaneously in conjunction with a collar locator in a cased or uncased hole.

record section *n:* a cross section of the earth generated by computer from tapes that have recorded the sound vibrations reflected during seismic exploration. Expert interpretation of record sections can reveal what may be a trap for petroleum. Also called seismic section.

reservoir *n:* a subsurface, porous, permeable rock body in which oil and/or gas is stored. Most reservoir rocks are limestones, dolomites, sandstones, or a combination of these. The three basic types of hydrocarbon reservoirs are oil, gas, and condensate. An oil reservoir generally contains three fluids—gas, oil, and water—with oil the dominant product. In the typical oil reservoir, these fluids occur in different phases because of the variance in their gravities. Gas, the lightest, occupies the upper part of the reservoir rocks; water, the lower part; and oil, the intermediate section. In addition to its occurrence as a cap or in solution, gas may accumulate independently of the oil; if so, the reservoir is called a gas reservoir. Associated with the gas, in most instances, are salt water and some oil. In a condensate reservoir, the hydrocarbons may exist as a gas, but, when brought to the surface, some of the heavier ones condense to a liquid.

reservoir drive mechanism *n:* the process in which reservoir fluids are caused to flow out of the reservoir rock and into a wellbore by natural energy. Gas drives depend on the fact that, as the reservoir is produced, pressure is reduced, allowing the gas to expand and provide the driving energy. Water-drive reservoirs depend on water pressure to force the hydrocarbons out of the reservoir and into the wellbore.

rig *n:* the derrick or mast, drawworks, and attendant surface equipment of a drilling or workover unit.

rock *n:* an aggregate of different minerals. Rocks are divided into three groups on the basis of their mode of origin: igneous, metamorphic, and sedimentary.

rotary table *n:* the principal component of a rotary, or rotary machine, used to turn the drill stem and support the drilling assembly. It has a beveled gear arrangement to create the rotational motion and an opening into which bushings are fitted to drive and support the drilling assembly.

S

salt water *n:* a water that contains a large quantity of salt; brine.

saltwater disposal *n:* the method and system for the disposal of salt water produced with crude oil. A typical system is composed of collection centers (in which salt water from several wells is gathered), a central treating plant (in which salt water is conditioned to remove scale- or corrosion-forming substances), and disposal wells (in which treated salt waste is injected into a suitable formation).

samples *n pl:* 1. the well cuttings obtained at designated footage intervals during drilling. From an examination of these cuttings, the geologist determines the type of rock and formations being drilled and estimates oil and gas content. 2. small quantities of well fluids obtained for analysis.

sampling *n:* the taking of a representative sample of fluid from a tank to measure its temperature, specific gravity, and BS&W content.

scale *n:* a mineral deposit that precipitates out of water and adheres to the inside of pipes, heaters, and other equipment.

screen liner *n:* a pipe that is perforated and arranged with a wire wrapping to act as a sieve to prevent or minimize the entry of sand particles into the wellbore. Also called a screen pipe.

secondary recovery *n:* 1. pressure maintenance methods used to boost primary reduction. 2. waterflooding of a depleted reservoir. 3. the first enhanced recovery method of any type applied to a reservoir to produce oil not recoverable by primary recovery methods. See *pressure maintenance, enhanced oil recovery,* and *primary recovery.*

sediment *n:* 1. the matter that settles to the bottom of a liquid; also called tank bottoms, basic sediment, and so forth. 2. in geology, buried layers of sedimentary rocks.

sediment and water *n:* a material coexisting with, yet foreign to, petroleum liquid and requiring a separate measurement for reasons that include sales accounting. This foreign material includes free water and sediment (dynamic measurement) and/or emulsified or suspended water and sediment (static measurement). The quantity of suspended material present is determined by a centrifuge or laboratory testing of a sample of petroleum liquid.

seismic data *n:* detailed information obtained from earth vibration produced naturally or artificially (as in geophysical prospecting).

seismic section *n:* Also called record section. See *record section.*

seismograph *n:* a device that detects vibrations in the earth, used in prospecting for probable oil-bearing structures. Vibrations are created by discharging explosives in shallow boreholes, by striking the surface with a heavy blow, or by vibrating a heavy plate in contact with the ground. The type of velocity of the vibrations as recorded by the seismograph indicate the general characteristics of the section of earth through which the vibrations pass.

separator *n:* a cylindrical or spherical vessel used to isolate the components in streams of mixed fluids. See *oil and gas separator.*

shale *n:* a fine-grained sedimentary rock composed of consolidated silt and clay or mud. Shale is the most frequently occurring sedimentary rock.

shale oil *n:* See *oil shale.*

shale shaker *n:* a vibrating screen used to remove cuttings from the circulating fluid in rotary drilling operations. The size of the openings in the screen should be carefully selected to be the smallest size possible that will allow 100 percent flow of the fluid. Also called a shaker.

shaped charge *n:* a relatively small container of high explosive that is loaded into a perforating gun. Upon detonation, the charge releases a small, high-velocity stream of particles (a jet) that penetrates the casing, cement, and formation. See *perforating gun.*

shut-in bottomhole pressure *n:* the pressure at the bottom of a well when the surface valves on the well are completely closed. The pressure is caused by fluids that exist in the formation at the bottom of the well.

shut-in bottomhole pressure test *n:* a bottomhole pressure test that measures pressure after the well has been shut in for a specified period of time. See *bottomhole pressure test.*

slips *n pl:* wedge-shaped pieces of metal with teeth or other gripping elements that are used to prevent pipe from slipping down into the hole or to hold pipe in place. Rotary slips fit around the drill pipe and wedge against the master bushing to support the pipe. Power slips are pneumatically or hydraulically actuated devices that allow the crew to dispense with the manual handling of slips when making a connection. Packers and other downhole equipment are secured in position by slips that engage the inner surface of casing.

solvent *n:* a substance, usually liquid, in which another substance dissolves.

sour crude *n:* also called sour crude oil. See *sour crude oil.*

sour crude oil *n:* oil containing hydrogen sulfide or another acid gas.

sour gas *n:* gas containing an appreciable quantity of hydrogen sulfide and/or mercaptans.

specific gravity *n:* the ratio of the weight of a given volume of a substance at a given temperature to the weight of an equal volume of a standard substance at the same temperature. For example, if 1 cubic inch of water at 39°F weighs 1 unit and 1 cubic inch of another solid or liquid at 39° weighs 0.95 unit, then the specific gravity of the substance is 0.95. In determining the specific gravity of gases, the comparison is made with the standard of air or hydrogen.

speed reducer *n:* a set of gears installed between a prime mover and the equipment it drives to reduce the running speed. For example, on a beam pumping unit, the engine may run at a speed of 600 revolutions per minute but the pumping unit it drives may need to operate at 20 strokes per minute. The speed reducer makes it possible to obtain the correct pump speed.

spherical separator *n:* a separator that is a round, ball-shaped vessel. See *separator.*

standing valve *n:* a fixed ball-and-seat valve at the lower end of the working barrel of a sucker rod pump. The standing valve and its cage do not move as does the traveling valve.

steam drive *n:* a method of enhanced oil recovery in which steam is injected into a reservoir through injection wells and driven toward producing wells. The steam reduces the viscosity of crude oil, causing it to flow more freely. The heat vaporizes lighter hydrocarbons; as they move ahead of the steam, they cool and condense into liquids that dissolve and displace crude oil. The steam provides additional gas drive. This method is used to recover viscous oils. Also called continuous steam injection or steam flooding.

steam flooding *n:* See *steam drive.*

stick *n:* a solid corrosion inhibitor that can be dropped down a well where it dissolves and mixes with the well fluids. See *corrosion inhibitor.*

stimulation *n:* any process undertaken to enlarge oil channels or create new ones in the producing formation of a well (e.g., acidizing or formation fracturing).

stock tank *n:* a storage tank for treated crude oil. Compare *production tank.*

storage tank *n:* a tank in which oil is stored pending transfer by pipeline, truck, or other vehicle for selling.

strata *n pl:* distinct, usually parallel beds of rock. An individual bed is a stratum.

stuffing box *n:* a device that prevents leakage along a piston rod, propeller shaft, or other moving part that passes through a hole in a cylinder or vessel. It consists of a box or chamber made by enlarging the hole and a gland containing compressed packing. On a well being artificially lifted by means of a sucker rod pump, the polished rod operates through a stuffing box, preventing escape of oil and diverting it into a side outlet to which is connected the flow line leading to the oil and gas separator or to the field storage tank. For a bottomhole pressure test, the wireline goes through a stuffing box and lubricator, allowing the gauge to be raised and lowered against the well pressure. The lubricator provides a pressure-tight grease seal in the stuffing box.

submersible pump *n:* a pump that is placed below the level of fluid in a well. It is usually driven by an electric motor and consists of a series of rotating blades that impart centrifugal motion to lift the fluid to the surface.

subsurface *adj:* below the surface of the earth (e.g., sursurface rocks).

subsurface safety valve *n:* See *tubing safety valve.*

sucker rod *n:* a special steel pumping rod. Several rods screwed together make up the mechanical link from the beam pumping unit on the surface to the sucker rod pump at the bottom of a well. Sucker rods are threaded on each end and manufactured to dimension standards and metal specifications set by the petroleum industry. Lengths are 25 or 30 feet (7.6 or 9.1 m); diameter varies from ½ to 1⅛ inches (12 to 30 mm). There is also a continuous sucker rod (trade name: Corod).

sucker rod pump *n:* the downhole assembly used to lift fluid to the surface by the reciprocating action of the sucker rod string. Basic components are barrel, plunger, valves, and hold-down. Two types of sucker rod pumps are the tubing pump, in which the barrel is attached to the tubing, and the rod, or insert, pump, which is run into the well as a complete unit.

sucker rod pumping *n:* a method of artificial lift in which a subsurface pump located at or near the bottom of the well and connected to a string of sucker rods is used to lift the well fluid to the surface. The weight of the rod string and fluid is counterbalanced by weights attached to a reciprocating beam or to the crank member of a beam pumping unit or by air pressure in a cylinder attached to the beam.

surface casing *n:* also called surface pipe. See *surface pipe.*

surface pipe *n:* the first string of casing (after the conductor pipe) that is set in a well, varying in length from a few hundred to several thousand feet. Some states require a minimum length to protect freshwater sands. Compare *conductor pipe.*

surfactant *n:* a substance that affects the properties of the surface of a liquid or solid by concentrating on the surface layer. Surfactants are useful in that their use can ensure that the surface of one substance or object is in thorough contact with the surface of another substance.

surge tank *n:* a tank or vessel through which liquids or gases are passed to ensure steady flow and eliminate pressure surges.

syncline *n:* a downward, trough-shaped configuration of folded, stratified rocks. Compare *anticline.*

T

tank *n:* a metal, plastic, or wooden container used to store a liquid. Three types include mud tanks for drilling, production tanks, and storage tanks.

tank battery *n:* a group of production tanks located in the field to store crude oil.

tertiary recovery *n:* 1. the use of advanced enhanced oil recovery methods that not only restore formation pressure but also improve displacement of oil by overcoming forces that keep the oil trapped in rock pores. 2. the use of any enhanced oil recovery method to remove additional oil after secondary recovery. See *secondary recovery* and *enhanced oil recovery.*

test separator *n:* an oil and gas separator that is used to separate relatively small quantities of oil and gas, which are diverted through the testing devices on a lease.

thermal recovery *n:* a method of enhanced oil recovery in which heat is introduced into a reservoir to lower the viscosity of heavy oils and facilitate their flow into producing wells. The pay zone may be heated by injecting steam (steam drive) or by injecting air and burning a portion of the oil in place (in situ combustion). See *steam drive* and *in situ combustion.*

thief *n:* a device that is lowered into a tank to take an oil sample at any desired depth. The sample will subsequently be used to determine the BS&W content of the oil in the tank. *v:* to obtain a sample of oil from a tank.

thief hatch *n:* an opening with a lid on top of a stock tank; used for measuring the oil level and for taking samples with a thief. See *thief.*

three-phase separator *n:* a separator that separates well fluids into oil or emulsion, gas, and water. See *separator.*

trap *n:* layers of buried rock strata that are arranged so that petroleum accumulates in them.

traveling valve *n:* one of the two valves in a sucker rod pumping system. The traveling valve moves with the movement of the sucker rod string. On the upstroke, the ball member of the valve is seated, supporting the fluid load. On the downstroke, the ball is unseated, allowing fluid to enter into the production column. Compare *standing valve.*

tray *n:* a horizontal device in a tower that holds liquid and provides a vapor-liquid contact step. Several common types are bubble-cap, perforated, or valve trays.

treat *v:* to subject a substance to a process or to a chemical reagent to improve its quality or remove a contaminant.

tube bundle *n:* the inner piping of a condenser or heat exchanger, typically consisting of a group of pipes placed inside the shell of a tank.

tubing *n:* small-diameter pipe that is run into a well to serve as a conduit for the passage of oil and gas to the surface.

tubing coupling *n:* a special connector used to connect lengths of tubing.

tubing hanger *n:* an arrangement of slips, built into a steel housing and engaged in the upper end of the wellhead, that serves as a support for the suspended tubing string.

tubing head *n:* a flanged fitting that supports the tubing string, seals off pressure between the casing and the outside of the tubing, and provides a connection that supports the Christmas tree.

tubing pump *n:* a sucker rod pump in which the barrel is attached to the tubing. See *sucker rod pump.*

tubing safety valve *n:* a device installed in the tubing string of a producing well to shut in the flow of production if the flow exceeds a preset rate. Tubing safety valves are widely used in offshore wells to prevent pollution if the wellhead fails for any reason.

tubingless completion *n:* a method of producing a well in which only small-diameter production casing is set through the pay zone, with no tubing or inner production string used to bring formation fluids to the surface. This type of completion has limited application in less-pressure, dry-gas reservoirs.

two-phase separator *n:* a separator that separates the well fluids into liquid and gas. The liquid (oil, emulsion, or water) goes out the bottom; the gas goes out the top. See *separator.*

U

uncased hole *n:* also called open hole. See *open hole.*

unconformity *n:* 1. lack of continuity in deposition between rock strata in contact with one another, corresponding to a gap in the stratigraphic record.

2. the surface of contact between rock beds in which there is a discontinuity in the ages of rocks.

V

valve *n:* a device used to control the rate of flow in a line, to open or shut off a line completely, or to serve as an automatic or semiautomatic safety device. Those with extensive usage include the gate valve, plug valve, globe valve, needle valve, check valve, and pressure relief valve.

vertical heater-treater *n:* a heater-treater whose cylinder stands upright, perpendicular to the ground. See *heater-treater.*

vertical separator *n:* a separator of cylindrical shape standing upright, perpendicular to the ground. See *separator.*

viscosity *n:* a measure of the resistance of a liquid to flow. Resistance is brought about by the internal friction resulting from the combined effects of cohesion and adhesion. The viscosity of petroleum products is commonly expressed in terms of the time required for a specific volume of the liquid to flow through an orifice of a specific size.

viscous *adj:* having a high resistance to flow.

W

walking beam *n:* the horizontal steel member of a beam pumping unit, having rocking or reciprocating motion.

water drive *n:* the reservoir drive mechanism in which oil is produced by the invasion of underlying water, which forces the oil into the wellbore. In general, there are two types of water drive: bottom-water drive, in which the oil is totally underlain by water, and edgewater drive, in which only the edge of the oil is in contact with the water. Bottom-water drive is more efficient.

waterflooding *n:* a method of enhanced oil recovery in which water is injected into a reservoir to remove additional quantities of oil that have been left behind after primary recovery. Usually, waterflooding involves the injection of water through wells specially set up for water injection and the removal of water and oil from production wells drilled adjacent to the injection wells.

water injection pump *n:* pump used in waterflooding or in water disposal. In waterflooding, the pump injects water into a well drilled in one part of a reservoir in order to force oil to a production well. In water disposal, the pump injects water produced from a well into a reservoir for disposal.

water injection well *n:* a well into which water is pumped in order to dispose of the water or to recover oil in a waterflood.

well *n:* the hole made by the drilling bit, which can be open, cased, or both. Also called wellbore, borehole, or hole.

wellbore *n:* a borehole; the hole drilled by the bit. A wellbore may have casing in it or it may be open (uncased); or a portion of it may be cased, and a portion of it may be open. Also called borehole or hole.

well completion *n:* the activities and methods necessary to prepare a well for the production of oil and gas; the method by which a flow line for hydrocarbons is established between the reservoir and the surface. The method of well completion used by the operator depends on the individual characteristics of the producing formation or formations. Such techniques include open-hole completions, sand-exclusion completions, tubingless completions, and miniaturized completions.

well fluid *n:* the fluid, usually a combination of gas, oil, water, and suspended sediment, that comes out of a reservoir. Also called well stream.

wellhead *n:* the equipment installed at the surface of the wellbore. A wellhead includes such equipment as the casinghead and tubing head. *adj:* pertaining to the wellhead (e.g., wellhead pressure).

well logging *n:* the recording of information about subsurface geologic formations. Logging includes records kept by the driller and records of mud and cutting analyses, core analysis, drill stem tests, and electric, acoustic, and radioactivity procedures.

well stimulation *n:* any of several operations used to increase the production of a well. See *acidize* and *formation fracturing.*

well stream *n:* See well fluid.

wet gas *n:* 1. a gas containing water, or a gas that has not been dehydrated. 2. also called rich gas.

wildcat *n:* 1. a well drilled in an area where no oil or gas production exists. With present-day exploration methods and equipment, about one wildcat out of every seven proves to be productive, although not necessarily profitable. 2. (nautical) the geared sheave of a windlass used to pull anchor chain. *v:* to drill wildcat wells.

winch *n:* a machine that pulls or hoists by winding a cable around a spool.

wireline *n:* a small-diameter metal line used in wireline operations; also called slick line.

wireline stuffing box and lubricator *n:* a stuffing box and lubricator used with wireline during a bottomhole pressure test. See *stuffing box* and *lubricator.*

wire rope *n:* a cable composed of steel wires twisted around a central core of fiber or steel wire to create a rope of great strength and considerable flexibility. Wire rope is used as drilling line (in rotary and cable-tool rigs), coring line, servicing line, winch line, and so on. It is often called cable or wireline; however, wireline is a single, slender metal rod, usually very flexible. Compare *wireline.*

wire-wrapped screen *n:* a relatively short length of pipe that has openings in its sides and a specially shaped wire wrapped around the pipe. It is used in wire-wrapped screen completions, usually in conjunction with a gravel pack. The screen and gravel pack block out sand and allow fluids to flow into the well through the openings in the screen.